ORGANIC ADDITIVES AND CERAMIC PROCESSING, SECOND EDITION

With Applications in
Powder Metallurgy, Ink, and Paint

ORGANIC ADDITIVES AND CERAMIC PROCESSING, SECOND EDITION
With Applications in Powder Metallurgy, Ink, and Paint

by

Daniel J. Shanefield
Rutgers University

KLUWER ADADEMIC PUBLISHERS
Boston / Dordrecht / London

Distributors for North America:
Kluwer Academic Publishers
101 Philip Drive
Assinippi Park
Norwell, Massachusetts 02061 USA

Distributors for all other countries:
Kluwer Academic Publishers Group
Distribution Centre
Post Office Box 322
3300 AH Dordrecht, THE NETHERLANDS

Library of Congress Cataloging-in-Publication Data

A C.I.P. Catalogue record for this book is available
from the Library of Congress.

Printed on acid-free paper.

Printed in the United States of America

Contents

PREFACE   xi

1. INTRODUCTION   1

  1.1 Historical Overview   1

    Plasticity in Ancient and Modern Processing   1

    Specialized Functions of Additives   3

    Organic versus Inorganic   3

  1.2 Typical Organic Additives   4

2. CHEMICAL AND PHYSICAL BONDING   8

  2.1 Ordinary Bond Types   8

    Electron Pair Orbitals   8

    Ionic Bonds   10

    Covalent Bonds   10

    Physical Bonds   11

  2.2 Lewis Acids And Bases   12

  2.3 Hydrogen Bonds   14

  2.4 Polarity   17

  2.5 Radicals   18

  2.6 Typical Structures   19

3. ORGANIC CHEMISTRY FUNDAMENTALS   23

  3.1 The Naming System   23

    Hydrocarbons   24

    Alkyls   25

    Alcohols   31

    Carbonyl Compounds   33

    Carboxylic Acids   35

Cyclohexane and Related Structures   36
Other Positional Terms   41
Some Special Environmental Considerations   44
Cyclic Compounds With Double Bonds   47
Benzene Ring Compounds   49
Nitrogen Compounds   55
3.2 Reactions   56
A Few Important Reactions   56
Polymerization   59
3.3 Synthetic Polymers   62
Polyvinyls and Similar Thermoplastics   62
Thermosetting Polymers   70
High Performance Polymers   72
Copolymers   76
Characterization   76
3.4 Natural Carbohydrates   80
3.5 Hydrogen Bonds and Life   89

4. CERAMIC PROCESSING FUNDAMENTALS   91
4.1 The Process Steps From Powder To Ceramic 91
The Powder   92
The Slip   92
The First Shrinkage   94
Shaping the Green Body   95
Sintering   100
The Fired Ceramic Product   107
4.2 Agglomerates   109
The Causes of Agglomeration   109
Types of Agglomerates   110
4.3 Optimum Surface Area   112

5. PARTICLE CHARACTERISTICS — 115

5.1 The Packing Of Powders — 115
The Sizes Needed for Closest Packing — 115
The Sizes of Real Ceramic Powders — 120
The Bell Shaped Curve — 121
The Lognormal Distribution — 123
Additional Porosity — 124
Modified Particle Size Distributions — 125

5.2 Surface Area Calculations — 126

6. COLLOID SCIENCE, AS APPLIED TO CERAMICS — 131

6.1 Adsorption — 131
Bonding of the Adsorbate to the Powder — 131
Specificity — 136
Hydration of the Surface — 139

6.2 Charged Particles In Suspension — 140
Sources of Charge — 140
The Double Layer — 142
Measurement of the Charge — 148
Changing the Charge — 150

6.3 Stabilized Suspensions — 153
Charge Repulsion — 153
Steric Hindrance — 155

6.4 Viscosity — 158

6.5 Wetting — 168

7. SOLVENTS — 171

7.1 Predicting Solubility — 172
Nonpolar Materials — 172
Materials of Low Polarity — 172
Highly Polar Materials — 172
Polymers — 174

7.2 Hydrogen Bonding Effects 176

Achieving High Solids Loading 176

Bubbles and Foam 177

Evaporation Rates for Fast Drying Solvents 178

Slow Drying Solvents 181

7.3 Safety 184

Flammability 184

Toxicity Regarding Dosage 186

Toxicity Regarding Animal Tests 191

Toxicity Regarding Human Exposure 194

Random Variations in Toxicity Studies 196

Use First, Test Later 200

Potential Damage to the Broader Environment 202

7.4 Cost 207

7.5 Chemical Attack on the Powder 209

8. DISPERSANTS AND OTHER SURFACTANTS 211

8.1 Tests For Effectiveness 211

Sedimentation Height 211

Minimum Viscosity 212

Maximum Solids Loading at Maximum Viscosity 213

Maximum Green Density 215

8.2 Commonly Used Detergents 218

Anionic Surfactants 218

Cationic Surfactants 221

Nonionic Surfactants 224

8.3 Inorganic Surfactants 226

8.4 Organic Deflocculants For Ceramics 229

Aqueous Systems 229

Nonaqueous Systems 236

High Solids Loadings 251

9. BINDERS 255

  9.1 Burnout 259

    General Considerations 259

    Burning 261

    Evaporation 264

    Other Removal Methods 266

    Nonuniform Burnout 267

  9.2 Adhesion 269

  9.3 Green Strength 271

    Plasticizers 273

  9.4 Other Additives 275

10. PROCESSING EXAMPLES 280

  10.1 Dry Pressing 281

  10.2 Injection Molding 283

  10.3 Extrusion 284

  10.4 Tape Casting 286

  10.5 Slip Casting 290

APPENDIX I. GLOSSARY OF CERAMICS WORDS 291

APPENDIX II. WORDS USED IN COLLOID SCIENCE 297

APPENDIX III. INFORMATION SOURCES 302

INDEX 309

Preface To The Second Edition

This volume is intended to be used as a textbook for teaching purposes and also as a reference source for working engineers. Therefore, a wide range of subject matter must be covered, starting with fundamental explanations for students, and extending to advanced applications for development workers and factory problem-solvers. Such an ambitious task is being attempted only because of the present lack of resources which might otherwise fill the need.

The author planned the book for use as the primary text in an undergraduate course in processing, which he teaches at Rutgers University. However, the book could also be used as a supplementary text for more general courses in related subjects.

Powder metallurgy, printing inks, and paints involve many of the same organic additives as ceramic processing. These specialized fields of technology are usually covered somewhat by very general college courses in metallurgy, materials science, and chemical engineering, but there appears to be a need for more specific training in the area of the organic additives used in those fields. The formulators, for lack of confidence and better understanding, often rely on simple waxes or acrylates, when a higher level of technological knowledge could provide improved results. It is intended that this book will be useful as a supplementary source of information for those fields also, both as a self-teaching tool and for college coursework.

Courses covering the subject matter in this book are offered at institutions of continuing education for working engineers, often in conjunction with professional societies. The students in such courses, in addition to trained engineers, include suppliers of materials to the industries, market researchers, investors, and others who are peripheral to the industries themselves. A continuing education course with the same title as this book was taught by Dr. George Y. Onoda and the author starting in 1976. With periodic minor revisions, this course was then taught by the author and other colleagues, twice each year up to the present time, under the auspices of the Center for Professional Advancement of East Brunswick, NJ, USA and of Amsterdam, The Netherlands. In 1986, the author ended his 30 years in industry (spent principally at AT&T) to accept a post as Distinguished Professor of Ceramics Engineering at Rutgers University. He then began a college course on this subject, which can be attended by either undergraduates or graduate students. Notes from these various courses have been the skeleton of this book.

Because there are so few schools in which students can specialize in ceramics, many readers of this book (such as suppliers of additives) are not likely to be knowledgeable in the fundamentals of ceramic science. Therefore some attention is given to these principles, particularly where they are the foundation for understanding the usage of the organic additives. On the other hand, many specialists in ceramics engineering have not been trained in the fundamentals of organic chemistry. For that reason some attention is paid here to organic chemistry concepts and technology. A logically configured pyramid of simply-defined scientific words has been included throughout the text and appendices. This extends downward to common words such as "plastic" and "fluid." The extra attention to the explanations of simple terminology results from the author's recent teaching experiences with students in need of some remedial scientific training, and with students from countries where English is not the main language spoken.

Readers should be able to copy formulations from this book, go to the laboratory, and immediately make a workable ceramic product, at least as a running start for further developments. Case studies regarding specific problems from the author's industrial experience are presented, as well as decision tables that might help solve future problems. Regarding the future, many former students have told the author that the theoretical knowledge gained from the lectures became more important than the practical details, after time lapses of several years. This was because a rather deep understanding of *general* theory was indeed required to create new things, especially after the *specific* practical details were "used up," to put it in the words of one former student. Therefore this book does not hesitate to present considerable theory, in some cases penetrating down to the depth of the "philosophical," especially in subject areas such as quantum mechanics and randomness. One of the author's teaching techniques is to link these theory discussions to practical examples, either in ceramics or in everyday life. This is done in order to make the lessons more understandable and also more easily remembered over the long term. Another technique used in this book is to break down complex ideas into small, "bite-sized" pieces.

The ancient technology of ceramics uses many words in highly specialized ways. Indeed, some of these words such as "blunger," "pug mill," and "fettler" are hardly encountered at all outside of the fraternity/sorority of ceramists. Throughout this text, a word or phrase which is likely to be unknown to the novice reader is usually enclosed in quotation marks. Often these words are presented again in later discussions and need special emphasis, and such words or phrases are italicized the next time they appear. Sometimes they will even be italicized more than once, to provide the student with a perspective regarding the context of the rest of the paragraph, and also to help the student remember the meaning. (Some of the writing is intentionally repetitious,[*] for the same reasons.)

Many descriptions of scientific points in this text will be supported by footnotes referring to published literature. These references are meant to be typical articles on the subjects being introduced to the reader, but they are not a complete or comprehensive listing. Often there will be one or more old articles included for the historic record, if an important discovery is involved. The newer articles will usually have their own listings of the pertinent literature of intermediate age, so such intermediate articles are sometimes not included in the present listing. However, there will be exceptions to this, where the listing of special articles of intermediate age would greatly aid the novice reader in finding material for further study. Also, in some cases several articles cover quite different aspects of the same subject, in which cases they might all be included in the list.

For brevity, only the first author is listed in most cases. However, the other authors are also listed when they have each published additional work on the subject, so the reader can look for them easily in author indexes to research the subject in more detail. The first digits that appear in each literature reference are the volume number, the ones in

---

[*] For items of special interest to the reader, use of the index is encouraged, in order to link together different aspects of that item (for example, "imines").

parenthesis are the year, and the last digits are the beginning page number. (For suppliers of chemicals, see pages 302 and 306.)

Several of the items described in this book have not been well known to the ceramics community, and in fact some of these items might be getting their first public exposure here. It is the author's opinion that they are important and should receive more attention. These items include the widespread importance of hydrogen bonding in ceramic processing (page 14), the remarkable effectiveness of organometallic sintering aids (p. 107), the three-dimensional closest packed size distribution (p. 118), the "use it all" rule for riffled powder sampling (p. 128 footnote), the Lewis base nature of π electrons in adsorption (p. 133), a procedure for dispersing oils in water (p. 223), the need for steric hindrance dispersants to be mechanically "soft" materials (p. 238), techniques for obtaining very high solids loading (p. 235) and green density (p. 278), the effectiveness of oleyl alcohol as a dispersant (p. 253), liquid crystal superlubricants (p. 278), the use of an emulsion fog for external lubrication (p. 282), and some thinking techniques for solving industrial problems (see "logic" in the index).

Acknowledgments

The author was privileged to work with many technical assistants who contributed significantly to a large scale ceramics project at AT&T, and much of the information gained from that work is now summarized in this book. It is a pleasure to acknowledge the efforts of Matthew Andrejco, Gerald E. Crosby, Michael S. Gervasio, Terri Giverson, Norman E. Leaver, Joan Pendyke, William Schultz, Raymond ("Rick") E. Sinitski, and Lloyd E. Trego.

Special thanks are due to Dr. William H. Bauer, Professor Emeritus of Rutgers University, for first introducing the author to the field of ceramics. Dr. Bauer helped the author obtain his first professional job, at ITT Laboratories. Several decades years later, Dr. Bauer was instrumental in obtaining the author's present (and probably last) job, at Rutgers. Thanks are also due to Drs. Harold W. Stetson and Richard E. Mistler for teaching the author much that he now knows about ceramics, during many years as co-workers at AT&T. The opportunity to write this book was provided by the late Dr. Malcolm G. McLaren.

Princeton, NJ Daniel J. Shanefield

1. INTRODUCTION

1.1 Historical Overview

Plasticity in Ancient and Modern Processing. In the original, ancient technology of ceramics, a mixture of clay and water was found to have the useful characteristic of "plastic flow." This phrase means that the clay and water mixture can be made to flow into a desired shape by applying a moderate amount of force, for example, by the direct action of human hands. However, when the potter takes his or her hands away from the clay, and only the much smaller force of gravity is then present, the clay will no longer flow. Thus, while the plastic clay-water mixture can readily be shaped by moderate force, it will still keep this shape on the storage shelf until it is fired. Ceramic technology, ancient or modern, is hardly possible without this special capability. When wetted with water, some types of clay such as "ball clay" have better flow properties than others, and these are used for making ceramics where precise shaping is desired. (Note: many of these words are defined more completely in Appendix I.)

The word "processing" includes all the mixing, shaping, drying, and firing steps needed to take the starting materials to the end product condition. In the making of ancient pottery, the starting materials were simply clay (the "ceramic powder") and water (the "fluid"). In a modern process, the starting materials might be a man-made ("synthetic") ceramic powder such as aluminum oxide, plus a fluid such as alcohol, and many other things that are added to the mixture (the "additives"), such as polyethylene glycol, stearic acid, etc. The overall process might be quite complex, including ball milling, filtering, spray drying, dry pressing, and other steps prior to the usual firing. Many of these processing details also are used in the powder metallurgy, ink, and paint technologies.

In either an old or a new process that uses ball clay because of its excellent plastic flow property when it is wet, it is of great interest to scientifically understand exactly why this property exists, and how to control it and cause the flow to be tightly reproducible, both in summer and winter. Also, if a new truckload of ball clay comes into the ceramic factory, and it is somewhat different from the previous truckload, it might not have the right plastic flow properties. Therefore the ceramist (sometimes called the "ceramicist" in older literature) might have to change various things so that the new "lot" of starting material can be made usable. Possibly some new chemicals might need to be added. This is another reason why scientific understanding is useful.

Ceramics researchers have determined that the plastic nature of wet ball clay arises partly from approximately 0.5% of a chemical material, commonly called "humic acid" or "lignite," which naturally occurs in the clay, and which binds the particles together with just the right amount of strength. These chemicals are what remain after formerly-living materials such as wood or leaves have decayed during thousands of years of geological time. They are "organic" chemicals, meaning that they contain carbon, similar to the materials in living things. One can not buy a bottle of humic acid, since it is not a single chemical compound; instead, it is a mixture of many compounds, which varies somewhat randomly from sample to sample of ball clay. Quite a lot of information about it can be found in the literature referred to under "ball clay" in Appendix I of this book.

Modern ceramics often require the elimination of the natural clay and the use of a synthetic ceramic powder such as "alumina" (aluminum oxide) powder instead. This is done because often the synthetic powder offers greater strength (after processing and firing), higher temperature refractory performance, or better electrical behavior. An example of a non-clay ceramic is the white insulator in an automobile spark plug. About 50 years ago, these insulators were being made from natural clay-based porcelain, but now they are made from synthetic alumina starting materials because of the lower tendency to crack when stressed and also the higher electrical resistance.

The synthetic ceramic powders usually do not have sufficient plastic flow for practical shaping, since they do not contain the remains of decayed wood, leaves, etc. and they are sometimes called "nonplastics." Therefore specially chosen organic chemicals must be added, to provide plastic flow and other advantageous characteristics. In some formulations, the added organic chemicals are natural additives, such as starch. In other formulations, they are synthetic additives, such as polyvinyl butyral. Both are used in large quantities by the ceramics industry. Usually the synthetic additives are more expensive but have higher performance, quite analogous to the ceramic powders themselves.

To summarize at this point, either the ceramic powder (the main ingredient) or the additive (used in smaller quantities, but still important) can be natural or synthetic. They can be combined in many different ways, and each has certain advantages. Any one of these can be either organic or inorganic, and even the ceramic powder can be made from "organometallic precursors." A few examples are shown in Table 1.1. Many other examples will be discussed in later chapters. Some are used in the form of "solutions," some are "dispersions" of insoluble "particles," and some are the intermediate "sols." (Note: for further explanations of some of these words, the curious reader might turn briefly to Appendix II).

Specialized Functions of Additives. In the water-plus-clay mixtures of traditional ceramic compositions, the water itself has more than one simple purpose. While the mixture is being shaped (by human hands, for example), the water provides "fluidity," allowing easy plastic flow. After the shaping force is released and only gravity is present, the water helps the clay stick together more than dry clay alone would stick, so the water then provides a "binding" effect, adding strength. (The second point is similar to the case of damp sand being stronger and more easily maintained in a given shape than very dry sand, but sand does not flow as smoothly as damp clay.)

In modern ceramics, these two functions of water, fluidizing and binding, are often separated, and specialized materials can be used for each. Examples of the specialized functions are the use of alcohol to fluidize, and modified-cellulose to operate as a binder.

Literally hundreds of organic additives are used in the ceramics industry. Often these were originally selected by trial and error many years ago, without much scientific justification, and better materials are now available. The following summary attempts to place the typically-used additive materials in logical categories, and this listing will be expanded in more detail during later discussions. Also there will be more detail on processing, after certain conceptual building blocks in the field of chemistry have been presented.

Organic Versus Inorganic. Most organic chemicals, when compared to most inorganic chemicals, have the advantage of being burnable, so they can be completely removed during the early stages of firing. Therefore the formulator does not have to worry about residues that possibly would degrade the final product, and he or she can concentrate on optimizing the flowability and other properties during shaping. Thus the specialized additives are often organic chemicals. In contrast, inorganic chemicals are sometimes used instead, which do not contain carbon, and which do not burn out. These can be useful

Table 1.1 Typical Organic and Inorganic Additives for Ceramics

Purpose	Material		Advantages
	Aqueous	*Non-aqueous*	
Solvent	Water*[†]		Cheap, Safe
		Toluene + Ethanol	Fast Drying
		Trichloroethane	Non-flammable
Dispersant	Sodium Silicate*		Cheap
	NH4 Polyacrylate		No Residue
		Menhaden Fish Oil[†]	Wide Range
		Polyethyleneimine	Easy Burn-out
Binder	Starch[†]		Very Cheap
	Methyl Cellulose		Heat-Gelled
		Polyvinyl Butyral	Strong
		Acrylate Emulsion	Easy Burn-out
Plasticizer	Ethylene Glycol		Cheap
	Glycerin		Strong Effect
		Polyethylene Glycol	Cheap
		Octyl Phthalate	Release Agent
Lubricant	Wax Emulsion		Cheap
	NH4 Stearate		Homogeneous
		Wax[†]	Cheap
		Stearic Acid	Strong Effect

* Inorganic.   † Natural.

Additional Notes:

Each material often has more than one function. For example, some dispersants are also lubricants, and a binder is often a dispersant, etc.

Two or more similar materials are often used in the same composition. For example, two solvents mixed together are sometimes better than one.

For details of some practical compositions, see Chapter 10.

where the residues after firing are not detrimental. One example is that sodium silicate is acceptable for making ceramic tile, although it is not acceptable for making high temperature firebrick.

Because carbon readily bonds to itself, making a wide variety of slightly different chains, hundreds of thousands of distinctly different organic chemicals are available to us, including both natural materials and synthetics. This, in addition to easy burnout, is another advantage of organic additives: the enormous level of versatility that is available because of this unique degree of self-bonding in carbon compounds.

1.2 Typical Organic Additives

Solvents. The solvent is used to temporarily make the system more fluid. The binder, which is usually an organic chemical in solid form, is ordinarily dissolved in this solvent. The ceramic powder itself is not commonly dissolved, but instead it is put into the form of an "unstable suspension" (often called* a "slip") allowing easy flow. At a later stage in the process, the solvent is evaporated out of the system ("dried"), so it is really just a temporary material, used mostly for fluidizing. Note that the solvent fluidizes both the powder, which is not dissolved, and the binder, which usually is dissolved.

Water is the most commonly used solvent in ceramic technology. The main advantages of water are that it is cheap and safe. Water also has disadvantages, such as a tendency to chemically react with certain ceramic powders like barium titanate or aluminum nitride, degrading their properties. Also, water sometimes causes the slip to be too "viscous" (sometimes referred to as too "thick"). In addition, water does not evaporate as quickly as some organic solvents, and therefore drying is much slower and requires more heating.

For these reasons, organic solvents are often used instead of water. A typical example is toluene, which is flammable and also is suspected of being toxic. It is only used where water cannot be used for the above reasons. Obviously, then, the choice of solvent involves a trade-off between several advantages and disadvantages.

As mentioned above, the binder and some other components are usually dissolved in the solvent. Some binders will dissolve only in

---

* Other words for *suspension* or *slip*, which are used interchangeably in ceramics terminology, are *slurry* and *dispersion*. In ceramics processing, this is usually not a true "colloid" or "sol," because it is not completely "stable" in most cases — see Appendix II for more about these words.

water, and some others only in non-water ("nonaqueous") liquids such as toluene. The initial choice of the solvent therefore dictates the choice of chemical family for the other additives, also.[1] In general, the whole field of organic additives for ceramics can be divided conveniently into two categories of aqueous and nonaqueous chemical systems. Table 1.1 shows some of the alternative examples of solvents and other additives that are compatible with either the aqueous or nonaqueous solvents.

Dispersants. These materials, sometimes called "deflocculants" or "surfactants," are put into the composition for two main reasons: (1) to keep the powder-solvent slip from becoming too "viscous" (having too much resistance to flow), and (2) to prevent "agglomeration" (lumpiness). The use of a powerful dispersant such as a polyacrylate can provide a reasonably low viscosity, even when only small amounts of solvent are used. Also, freedom from the lumpiness of agglomerates will usually prevent large pores from being present in the material, which is another very critical consideration. Although dispersants represent only about 8% of the current market value[2] of the total usage of organic additives in the ceramics industry, this particular niche of the market is growing fast, because increased automation is requiring better reproducibility, and the newer dispersants are able to prevent the type of random variation in density that is caused by agglomerates.

Binders. After the ceramic shape ("body") has been molded and the solvent has been evaporated, the use of a binder can provide considerable strength, even before firing. Thus the unfired ("green") body is strong enough to be handled for inspection and for loading into the furnace. In some cases it can be machined on a lathe or drilled with precise holes, as is typically done with spark plug insulators, for example. Binders represent the largest segment of the current market for organic additives in ceramic processing.[2]

Plasticizers. In the older literature on ceramic technology, a "plasticizer" was a material such as ball clay which can provide plastic flow when it is wetted with water. The word is still sometimes used in that sense, but there is also another meaning. When this word is used in modern chemical technology, it means a liquid that can be added to the binder composition to prevent brittleness in the dry body. If the

[1] Y. Kankawa, "Organic Materials in the Production of New Ceramics," Seramik-kusu, 29 (1994) 585 [in Japanese].

[2] K. K. Chan and D. J. Shanefield, Am. Ceram. Soc. Bul., 68 (1989) 854.

binder is prevented from becoming brittle when it is dried, then it is effectively stronger, and the formulator can use less of it and still achieve a practical level of unfired green strength. An example of a plasticizer is glycerin, which is excellent for use with a polyvinyl alcohol binder.

Lubricants. The shaping ("forming") of modern ceramics is often done at very high applied pressures, such as in dry pressing or extrusion. The metal molds ("dies") will wear out quickly during this contact with very abrasive ceramic powder, unless a lubricant is used to decrease the wear. In addition, the lubricant can allow smoother flow of the powder under pressure, with fewer defects due to particles not flowing evenly. Stearic acid is a lubricant that is commonly used in ceramic processing. For water-based (aqueous) systems, it can be dissolved in the form of an ammonium stearate salt. For an organic solvent (nonaqueous) system, the stearic acid can be dissolved in alcohol.

Others. Several categories of organic additives such as release agents, antifoams, and biocides, which are of less importance, will be covered in later sections. However, those discussions will be more understandable after some further theoretical principles have been presented.

Of course, all of the organic materials briefly described above become "burned out" or else are evaporated out of the ceramic composition by the time the firing step is finished. This is one of the main advantages of organic versus inorganic additives, in addition to the seemingly unlimited variety of organic materials that can be designed by synthesizing chemists.

2. CHEMICAL AND PHYSICAL BONDING

2.1 Ordinary Bond Types

Electron Pair Orbitals. A practical understanding of ceramic processing requires some knowledge of chemical bonding. Weak bonds, strong bonds, and some bond types not ordinarily emphasized in first year chemistry courses are all involved in a typical ceramic process, even for an old method such as the slip casting of ball clay. An even deeper level of knowledge will be required in order to understand the functions (and occasional malfunctions) of those organic additives that are likely to be used in the future.

Bonding between two or more atoms is the essence of chemistry.[1] Figure 2.1 shows the approximate paths ("orbitals") of outer electrons around atoms which are bonded together. An **x** indicates an electron from one atom, and an **o** indicates an electron from the other atom, traveling in the circular or oval paths shown.[*] Two electrons which are spinning in opposite directions can pair together to form the type of orbital which forms a chemical bond. There is a precise set of natural rules that describes the behavior of small particles such as electrons, and these rules are a branch of physics called quantum mechanics. Within the past decade, quantum mechanics has been used to predict electron-pair orbitals for new chemical compounds, with a useful level of accuracy.

Whether or not being able to *predict* proves the *truth* of any particular theory is a matter for philosophers. Quantum mechanics is based on a rather mysterious[2] system of numerical relationships,

---

[1] J. E. Huheey, "Inorganic Chemistry," Harper, New York (1993).

[2] J. Horgan, "Quantum Philosophy: The Surreal Realm," Scientific American, Vol 267 (July 1992) page 94.

[*] The electrons actually occupy rather fuzzy areas, not the sharply-defined circles shown.

			Bond Strength	Molecule Polarity	Examples
PHYSICAL	van der Waals	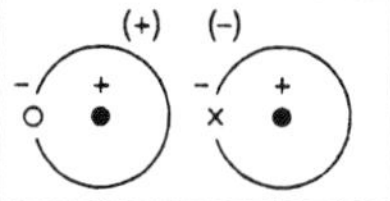 (+) (−)	Weak ~1 kcal/mol	Changing Polar	Graphite in C-direction
CHEMICAL	Ionic	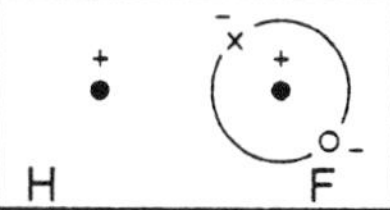 H F	Strong 100-250 kcal/mol	Constant Polar	HF
	Covalent	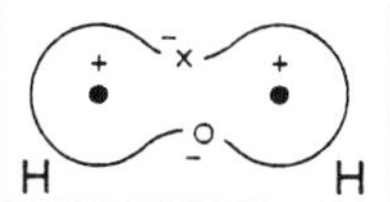 H H	Strong 50-150 kcal/mol	Nonpolar	H_2, epoxy
	Coordinate Covalent	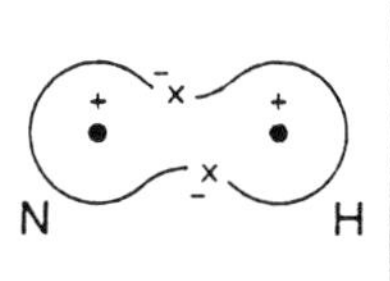 N H	Moderate ~50 kcal/mol	Nonpolar	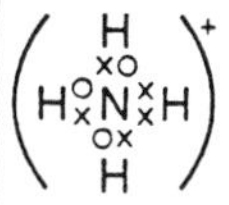 $\left(H_3N \rightarrow H\right)^+$

Fig. 2.1 Chemical bond types. It should be noted that, after a coordinate bond forms, it becomes nonpolar or only slightly polar, even though the two-atom situation might have been very polar before the bond had formed.

usually involving whole numbers rather than fractions. Certain numerical sequences appear to be favored by nature, and others are unlikely (called "forbidden" in quantum mechanics terminology). In some respects, quantum mechanics resembles ancient beliefs in numerology, in which 7 is lucky and 13 unlucky: we can not explain, past a certain point, why certain numbers (such as eight electrons in an energy shell) are favored by nature and not some other numbers. However, the modern science of quantum mechanics can predict the results of future experiments, while numerology, as far as the author knows, can not. Suffice it to say that quantum mechanics is a systematic description of bonding, and it has shown us many new ways to make improved chemical compounds.

Ionic Bonds. According to quantum mechanics, certain pairings of electrons are far more likely to occur than others, and these specific pairings result in the bonds between the atoms themselves. An example is the ionic bond shown by the simplified diagram in the figure. Although the "**o**" electron was originally supplied by the left-hand hydrogen atom in the diagram, and the "**x**" electron by the right-hand fluorine atom, after bonding they are completely shared around the fluorine and can not be experimentally distinguished from each other. The tendency to form a two-electron orbital is so strong that it constitutes an attraction force, and the formation of this bond can give off a hundred or so kilocalories of heat per mole of compound formed.

The energy given off is analogous to that of a stretched spring being allowed to relax to an unstretched condition. This released energy could be used to raise a weight, or do some other work. Possibly it would only be given off as heat, in other words, random motion of atoms. Thus, when two atoms attract each other, the bond that is formed gets shorter, like a stretched spring getting shorter, and there is an output of energy. Another analogy would be two magnets being allowed to attract each other, and the resulting energy output being used to raise a weight (via a rope and pulley arrangement), or produce frictional heat (via a partially engaged braking device).

In the case of two atoms, the rules of quantum mechanics show us that certain pairs of atoms will approach each other closely and form a *chemical bond*. This new bond would be *ionic* when one of the atoms involved attracts electrons much more strongly than the other, as shown in the figure.

Covalent Bonds. The *covalent* bond is similar except that its two-electron orbital extends around both atoms (again, see Fig. 2.1). Bonds are covalent when the two atoms each attract electrons to approximately the same degree. The hydrogen molecule is one example, shown by the simplified diagram in the figure. Another example is the bond between two carbon atoms, within the long molecule of a plastic material such as epoxy (not shown in the figure, but to be shown in Chapter 3). Still another example is the bond between silicon and carbon in silicon carbide (SiC) ceramic, where the silicon and carbon atoms attract electrons almost equally.

In a *coordinate* bond, the electrons are shared by the two atoms, but both electrons came originally from just one of these atoms. An example is the bond to the fourth hydrogen of the ammonium ion, as shown by the pair of **x** symbols in the lower right hand corner of the illustration. Both electrons came from the nitrogen atom, but they are then shared by the two atoms together.[†] That bond is quite important in causing organic chemical additives to adhere to ceramic powders, and it will be referred to many times again in this text. The arrow symbol is sometimes used for coordinate bonds, as shown in the figure.

Note that the ammonium ion as a whole has a net electric charge, which is positive, so the whole structure can make an ionic bond to another ion such as chloride, etc. But within the ion, the four bonds are covalent. Another point, to be remembered for later use, is that these particular hydrogens can be substituted by four *organic* groups, and this is the basic structure of the "quaternary ammonium salts" to be discussed in later chapters, particularly in the lubricants and biocides sections.

Physical Bonds. The *van der Waals* bonds (some aspects of which are called the London Force or the Dispersion Force) are so weak that they are not usually classified as true chemical bonds. Instead, they are within the broader category of *physical bonds*. However, these bonds can be important in ceramics. They arise from the fact that part of the time an electron will be at the back of an atom (shown at the upper left in Fig. 2.1), so the positively-charged nucleus will be exposed. This positive charge would attract the negatively-charged electron of a neighboring atom. However, at other times the two electrons will be facing each other and repel. The overall net result can be estimated using rather complex quantum mechanical calculations, and it turns out to be a slight attraction. (An analogy that is easy to imagine is two magnets floating on corks in a pool of water, with small waves that disturb the positions randomly. Most of the time the magnets will

---

[†] It should be noted that once these bonds have formed, they become indistinguishable, and an observer has no way to determine which electron originally came from which atom, or which bond is the coordinate one. Because of one of the rules of quantum mechanics, nature favors the situation where several different positions of the electrons can exist and they are all equivalent. This principle is called "resonance" and will be mentioned again in Chapter 3, in connection with the discussion on benzene rings.

orient themselves north-facing-south, so there will be a net attraction, although this will not always be the case. However, the analogy is not exact, since the rules of quantum mechanics modify the common sense result to some degree, decreasing the net attraction.)

The van der Waals bond is so weak that, in practice, solid surfaces must be very clean and closely mated for it to have an appreciable effect. Although it is weak, it can still be important, because it can form between the two surfaces of all known materials, and therefore it must be considered as a possibility in many different circumstances. It is usually the force that holds adhesive tape, glue, and liquids to solid surfaces. It is also the "cohesive" force that pulls molecules toward each other within many soft solids such as wax. It is often one of the forces that hold the binder to the ceramic powder in green ceramic bodies. It holds the binder to the powder in many powder metallurgy bodies, inks and paints.

2.2 Lewis Acids and Bases

Further consideration of coordinate bonding is necessary in order for some newly-discovered aspects of ceramic processing to be fully understood. The first item to be discussed is a concept of acids and bases that might be unfamiliar to many readers, even engineering students.

The most commonly used definition of an *acid*, the so-called *Bronsted-Lowry* definition, is "a material *from which* a *hydrogen ion* is easily obtained," and this is often observed when an ionic material is dissolved in a solvent such as water. Important as this is, there is another type of acid/base concept which is also useful in ceramics, but which is not always studied in first year chemistry courses. In this concept, a "Lewis acid" is defined[3,4] as "a material which tends to *attract electrons* from other atoms." By comparison, a "Lewis base" has "plentiful *electrons to donate* to other atoms." Thus, the definition is thought of in terms of electrons instead of hydrogen ions.

Looking at the example of hydrogen chloride (HCl), this whole molecule is a conventional Bronsted-Lowry acid, because it easily pro-

---

[3] W. B. Jensen, "The Lewis Acid-Base Theory," J.Wiley, New York (1980).

[4] P. C. Stair, Langmuir, 7 (1991) 2508.

vides hydrogen ions when dissolved in water. However, in the *Lewis acid/base* system, the whole HCl molecule is not especially acidic. Instead, just the hydrogen ion itself is the strong Lewis acid, because it attracts electrons.

Usually, an atom with *fewer than four* electrons of its own in the outer shell, such as hydrogen with one electron, or aluminum with three, can act as a *Lewis acid*. This is because aluminum bonded to three other atoms, such as three oxygens in alumina, will have three pairs of electrons in its outer shell (six electrons total). That would not complete the *octet* (the eight-electrons making up the four orbitals that quantum mechanics predicts[1] to be an unusually stable combination), and therefore more electrons will be attracted, making the aluminum act as a Lewis acid.

Usually an atom with *more than four* electrons in its outer shell, such as oxygen with six, or nitrogen with five, can act as a *Lewis base*. The extra pairs of electrons tend to become donated. This is what takes place when coordinate bonds are formed. An example is nitrogen in the ammonium ion, as shown in Fig. 2.1 and discussed above. The example of the Lewis acid concept in the figure is the fourth hydrogen (on the right), which has no electrons of its own and thus attracts the extra electron pair of the nitrogen. In other words, the Lewis acid/base concept can be thought of as a further analysis of the coordinate bond (although there are other aspects which will not be dealt with in this book).

In ceramic technology, an important example of the Lewis concept is the case where the dispersant "polyethyleneimine" is bonded to alumina powder. (The "imine" structure, to be described in a later section, contains nitrogen.) Here, aluminum acts as a Lewis acid (electron seeking), with nitrogen in the imine group acting as a Lewis base (electron donating). A weak but still important chemical bond forms, and it is of the coordinate type. This provides one of the best and most practical dispersion actions in modern ceramic processing.

A pair of atoms bonded together with a *carbon-carbon double bond* (to be discussed later in connection with organic chemical structures) can act as a Lewis base. The electrons in a double bond stick out prominently in what is called the π configuration, and these electrons are more readily available for donation than in ordinary single bonds. Each of these Lewis acid/base examples can be involved in providing dispersion (deflocculation) in ceramic slips. They can also be involved in providing adhesion of the binder to the ceramic powder, and the adhesion of lubricants to extrusion dies, tool bits, etc.

2.3 Hydrogen Bonds

A special type of coordinate bond which is very important to ceramists is the *hydrogen bond,* shown in Fig. 2.2. Both electrons come from the same atom, that is, the oxygen.[1] This bond type has been suggested[5,6] as one of the main causes of powder agglomeration, which can be a critical factor in ceramic processing. In many cases it is also involved in the adherence of organic additive molecules to powder surfaces, and in the solubilities of binders.

The **x** symbols shown in the diagram are meant to represent electrons supplied by the hydrogen atoms, and the small **o** symbols represent electrons supplied by the oxygen atoms. According to the octet rule, as mentioned earlier, the outer electrons of most atoms tend to fill in more and more electron pairs until four orbitals are present (eight total electrons), whenever this is possible. Thus the hydrogen will bond to the oxygen, because this is going in the direction of a more nearly completed outer shell of four orbitals around the hydrogen, even though it is not fully completed.

Other pairs of electrons (**o** symbols) around the oxygen are not involved in the main bonding of an individual water molecule. However, if the hydrogen of another water molecule approaches two of these non-bonding electrons on the oxygen atom, a coordinate bond could form, tending to go toward a completed shell around the hydrogen. Such a bond has formed between the two water molecules shown in the diagram at the upper right. It is identifiable as a definite chemical bond because of its infrared spectrum and heat of formation, and also by other means. These hydrogen bonds can extend in several different directions from each oxygen atom of a given molecule, making a three-dimensional network. In most cases, once the bond is

5 D. J. Shanefield, Matls. Res. Soc. Proc., 40 (1985) 69.

6 R. G. Horn, J. Am. Ceram. Soc., 73 (1990) 1117 [see page 1128, water effects].

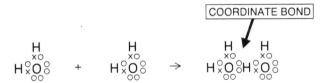

Two Molecules of Water Forming a Hydrogen Bond

Multiple Molecules of Water

Alcohol

Fig. 2.2 Hydrogen bonds, in water and alcohol.

formed, the hydrogen is shared equally by the two nearest oxygens, and the average bond strengths (and lengths) are just about equal on each side.

In liquid water, this can go on in more than one dimension and form a large network. If a chemist uses this definition of a "bond" (and there is no reason not to), then a glass of water is really all one molecule. In fact, since all the world's major oceans are connected together, via the Strait of Magellan below South America, etc., it is actually true that the whole world's oceans are one single molecule! (That might be something to remember, the next time the reader is asked, "About how big is a molecule?")

The hydrogen bonds (often called "H-bonds") are rather weak, so water can therefore flow by mechanically breaking down the H-bonds and re-formation of new ones whenever the opportunity arises. Alcohol, which also has OH ("hydroxyl") groups, can similarly form H-bonds. (In the figure, the "R" stands for "radical," which can be a methyl group, or an ethyl, or other type of organic chemical group that is part of the molecule, and this will be explained in later sections.) There are not as many opportunities for H-bonds to form in alcohol as there are in water, so the effects are not quite as strong.

If it were not for H-bonds, both water and methyl alcohol would be gases at room temperature as methane is. The H-bonds cause them to cohere as liquids, and this also raises the viscosity. Since the effect is greater in water, with more H-bond possibilities, its boiling point and viscosity are higher than those of alcohol.

Other materials that can easily form H-bonds, such as methyl cellulose, can be dissolved in water to purposely raise the viscosity. This is often done in modern mass produced foods that contain a lot of water, where methyl cellulose can be dissolved in that water, examples being mustard, mayonnaise, and ice cream. The H-bonds are weak enough to break easily, but they form again quickly between other pairs of molecules, thus providing fairly viscous flow. A high viscosity is also useful in the extrusion of ceramics, so that the freshly extruded shape does not collapse due to gravitational forces.

An important example in ceramic processing is in extrusion. In fact, the additive molecule ("methyl cellulose") can be formulated so that the H-bonding becomes even more effective at higher temperatures. With the use of such an additive, special heaters can be positioned immediately after the extruder, to prevent complex and delicate extruded shapes from collapsing. Without such modern technology, the honeycomb ceramic structures made for automobile catalytic converters could not be manufactured in a practical manner.

Because of the considerable importance of hydrogen bonds, readers who are pursuing research in this area might be interested in the fact that H-bonds are about 1.8 Angstroms (0.18 nm) long,[7] which is quite a lot longer than most covalent bonds, and this is one of the indications that they are relatively weak bonds. Absorption bands at 1920, 2500, and 3550 cm⁻¹ appear in the infrared spectrum‡ when H-bonds are present.[8]

2.4 Polarity

A chemical concept which is useful in understanding the details of what goes on in typical ceramic processing is "polarity." This can be defined as "the degree to which the electrons spend more time on one side of a chemical compound than they do on the other side." In other words, a *polar* molecule such as pure sulfuric acid (hydrogen sulfate, or H_2SO_4) has the electrons spending more time around the sulfate group than around the hydrogen atoms. Therefore the sulfate group has a slightly negative electric charge around it, and each of the two hydrogens is slightly positive.

If sulfuric acid is dissolved in water, it becomes even more polar, because the hydrogen (or at least some of it) becomes ionized with a positive charge, and some of the sulfate group is also ionized with either one or two negative charges. There can also be some degree of polarity even if the material is not at all ionized, an example being pure carbon monoxide, where the oxygen has slightly more negative charge around it than the carbon does. Thus, there can be a high degree of polarity, or a slight degree, or even a completely "nonpolar" material such as pure diamond. Many other examples of polarity will be

---

7 W.G. Lawrence, in G. Y. Onoda and L. L. Hench, "Ceramic Processing Before Firing," J. Wiley, New York (1978) 197.

8 C. H. Rochester, Adv. Colloid and Interface. Sci., 12)1980) 43 (see particularly pages 57, 62); J. A. Dean, "Lange's Handbk. of Chem." (1972) pages 8-64 and 8-72; M. D. Sacks, et al., Advances in Ceramics, 21 (1987) 500; J. Y. Lee, et al., Macromolecules, 21 (1988) 954; T. Kato, et al., Chem. Mat'ls., 5 (1993) 1098.

‡ Just as a piano string vibrates at a specific frequency depending on the weight, length, and tightness of each string, a chemical bond vibrates depending on the weights of the bonded atoms, and the length and stiffness of the bond. If light of the right frequency passes by, it will give up its energy (and thus be "absorbed") to make the bond vibrate. Thus each infrared absorption band corresponds to a specific type of bond.

discussed during the explanation of why certain materials dissolve in water or don't dissolve, and also in other chapters.

2.5 Radicals

Organic chemistry terminology makes frequent use of the word "radical," meaning an incomplete section of a molecule, or any "group" of atoms that are chemically bonded together but have an unsatisfied ("dangling") bond. In some types of diagrams it is symbolized by the capital letter R (as in Fig. 2.2). For example, the *radical* might be the "ethyl" group in "ethyl alcohol," which is shown in Fig. 3.3. (The words radical and group have almost the same meaning, but group is slightly more broad and general, since it is applied to inorganic materials and other items also.)

When a chemical bond between two atoms is broken, sometimes both of the electrons in the bonding orbital will stay with just one of the atoms, leaving the other atom with no electrons from that particular bond. One example is sulfuric acid that has been dissolved in water, where the sulfate ion holds onto all of the bonding-orbital electrons, and the hydrogen ion loses all of its electrons. The sulfur and oxygen atoms each contain several protons within their central nuclei, but they have obtained even more of these outer electrons, and thus the sulfate ion has a net negative charge. The hydrogen ion, on the other hand, has lost its electron and is thus a positively charged proton. (Actually, the hydrogen ion is strongly attached to the surrounding neutral water by coordinate H-bonding, but the net charge is nevertheless positive, and it is therefore a positive ion.)

It is useful to realize that a chemical bond can also break in another manner, particularly in various organic chemicals. One electron of the original two-electron bond can go with each atom, resulting in no net electric charge. (Inside the nucleus of the atom, there is a positively charged proton for each electron, neutralizing the total charge.) The single electron without any net charge can be thought of as a "dangling, unsatisfied bond." It is located on a chemical group which is called a *free radical.* This is a radical which is not attached to anything and is therefore "free."

The importance of free radicals in everyday life has only recently been established, by the use of newly available instrumentation. For example, free radicals are probably an active part of the natural aging process in animals, and materials that can diminish the free radical concentrations have tended to prolong life in certain biological experiments. A few materials such as nitric oxide, NO, are *stable free radicals,* which can exist in spite of the unsatisfied, single electron dangling bond. A simplified structure can be drawn as ·N=O where the dot indicates a single, unpaired electron. Present evidence indicates that nitric oxide is important in many aspects of animal biology, especially some functions such as the erection of sex organs. Other free radicals are being mentioned more and more in the popular press and news media, in connection with cancer, human aging, environmental problems such as the ozone hole, etc.

Organic compounds such as binders for use in ceramics are often manufactured by chemical reactions that involve free radicals, in addition to the more widely known ionic reactions. On the other hand, when ceramists remove the organic binders during firing, these compounds sometimes break down through reactions which are the reverse of the ones that were originally used to manufacture them, and thus free radicals are also involved in the burnout steps.

2.6 Typical Structures

Figure 2.3 shows the several types of bonding in various materials. The van der Waals bonds and tangling are relatively weak, and materials can usually be made to flow or break in those particular directions without requiring much force. (However, in perfect single crystals such as graphite "whiskers," even the van der Waals bonds in one direction can be quite strong. In fact, they are sometimes stronger than the covalent bonds in ordinary materials which are imperfect and contain many defects.)

In polymers which require tangling for strength, longer molecules such as polyethylene are slightly stronger than shorter ones such as wax, because there is more tangling. The degree of strengthening due to tangling can be estimated by the science of "statistical mechanics." In some specially made polymeric materials, another factor can vastly increase the strength: crystallization. A new series of highly crys-

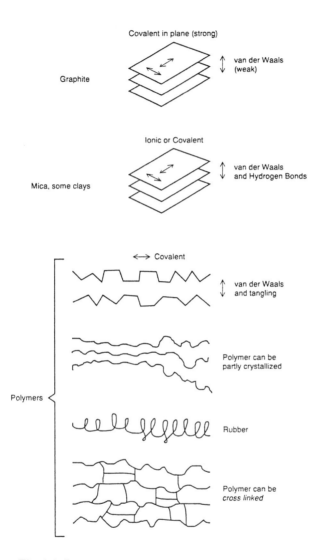

Fig. 2.3 Bonding in some solid materials. (Further discussion of the polymeric structures will appear in Chapter 3, Section 3.3.)

talline polyethylenes is extremely strong and is now being used in high performance applications such as ship cables. However, most polymers are just held together by van der Waals bonds and tangling and are not very crystalline. Therefore these are fairly soft materials having only moderate strength.

When a carbon atom has its usual four covalent bonds extending to other atoms, these bonds are arranged in *tetrahedral* fashion, as shown in Fig. 2.4. The dotted lines are not bonds but have simply been drawn

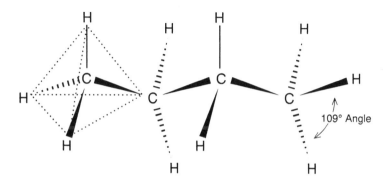

Fig. 2.4 The tetrahedral configuration of covalent bonds around the carbon atoms in butane.

to show the edges of an imaginary tetrahedron, with the carbon in the middle. It can be shown geometrically (and it has been verified experimentally) that any two of these bonds makes an angle of approximately 109°. The organic chemical compound shown in the figure is called *butane,* with four carbon atoms bonded to each other. Hydrogen atoms fill in the other potential bonding sites. This is a rather complex three-dimensional structure in which most of the hydrogen atoms are not actually in the plane of the paper used to make the diagram, with some of the hydrogens extending out toward the reader, and some bending back behind the plane of the paper. That fact is best illustrated by a stick-and-ball model, but the structure can be approximated fairly well on two-dimensional paper, using narrow triangles to symbolize bonds where the hydrogen atoms are toward the reader, and multiple dashes to symbolize bonds where the hydrogens are behind the carbons, away from the reader. Looking at any narrow triangle-shaped bond symbol, the larger end is toward the reader, in the manner of a perspective drawing. (Of course, the true chemical bonds are not triangles, but are fuzzy areas where electron pairs spend most of their time.) A simple, straight line represents a bond which is parallel to the plane of the paper. These drawing conventions are used more and more in patents and other chemical literature that the modern ceramist is likely to read.

Often the structures are simplified by completely "flattening" them, as will be seen on later pages. To make the jobs of teacher and student easier, most of the diagrams in this book have been flattened.

3. ORGANIC CHEMISTRY FUNDAMENTALS

3.1 The Naming System

Before attempting to deal with other aspects of organic chemistry, it is necessary that the nomenclature (naming system) of chemistry be understood. Many organic chemicals have more than one name, an example being a particular type of alcohol containing three carbon atoms, which can be called any of the following five names:

> "isopropyl alcohol," or "isopropanol," or
> "secondary propyl alcohol," or "sec-propanol," or
> "2-propanol."

The first example is the *common name,* which is part of an old system that is still frequently used. It is not completely logical or systematic in the way it puts together syllables to make up the complete name. Its first syllable, "iso-," might unfortunately mean any of several different things, when used to name some complex structures, but it is quite satisfactory for use with simple materials. On the other hand, the "2-" of the last name in the example "2-propanol" has a very clear meaning, even when the material is quite complex. Therefore, this has been chosen for use in the *systematic name* terminology. (These names will all be explained on later pages.) The other names in the example are special types of the common format, which are somewhat more systematic than the first one. They are not part of a completely logical naming system, so they are halfway between the common and systematic systems. It should be noted that capital letters are not ordinarily used in chemical names for either compounds or elements, although they are used at the beginnings of the short *symbols* for the elements (C for carbon, Br for Bromine, etc.). In chemical nomenclature, very often a long group of syllables is strung together with no break, an example being polyethyleneimine.

The following sequence of examples is designed to be an explanation of both the common and the systematic nomenclature systems of organic chemistry, with a few other intermediate naming types included wherever they are likely to be of use in ceramics. Further and more detailed information, along with a great deal of other interesting chemistry, is contained in a well-explained textbook by H. Hart.[1] Another source of information, which contains the complete scheme of international standard names for organic chemicals is the section on nomenclature in the CRC Handbook.[2]

Hydrocarbons. Figure 3.1 illustrates some of the simplest organic chemicals. For ease of explanation, the true three-dimensional nature of the structures is ignored, and instead they are shown in a simplified, "flattened" format. In many cases, both the structures and the names will be written several different ways in the illustrations, to accustom the reader to the many ways they are likely to be encountered later in working situations.

As illustrated in the figure, methane is the first member of the series, and it has only one carbon atom. More carbons can be added to the chain, making molecules that include thousands of atoms. As greater numbers of carbon atoms are added, the compounds tend to have higher densities, higher melting point temperatures, and higher boiling points. Thus, methane and ethane are gases (low densities, low boiling points), but octane is a liquid (having eight carbon atoms in the molecule, and purposely written somewhat differently in this particular example). The compound with 17 carbon atoms (heptadecane, not shown in the diagram) is a solid. Some longer-chain members of the series (not shown) are called waxes, which are solids. Even longer members are called "polyethylenes" (see page 62), and they are slightly stronger solids, because of the still greater tangling.

These trends and general principles will be noted again when the subject of *binders* for ceramic processing is discussed, where longer organic molecules are used to make stiffer, stronger solid binders, but also more brittle ones. If the brittleness gets to be a significant problem, a somewhat shorter organic molecule which is a liquid can be added to the formulation to act as a *plasticizer*, decreasing the tendency of the material to crack when it is handled during normal operations such as inspection, decorating, etc.

---

[1] H. Hart, "Organic Chemistry: A Short Course," Houghton Mifflin, Boston (1987).

[2] "Handbook of Chemistry and Physics," CRC Publishers, Boca Raton, FL (1993) page 2-45.

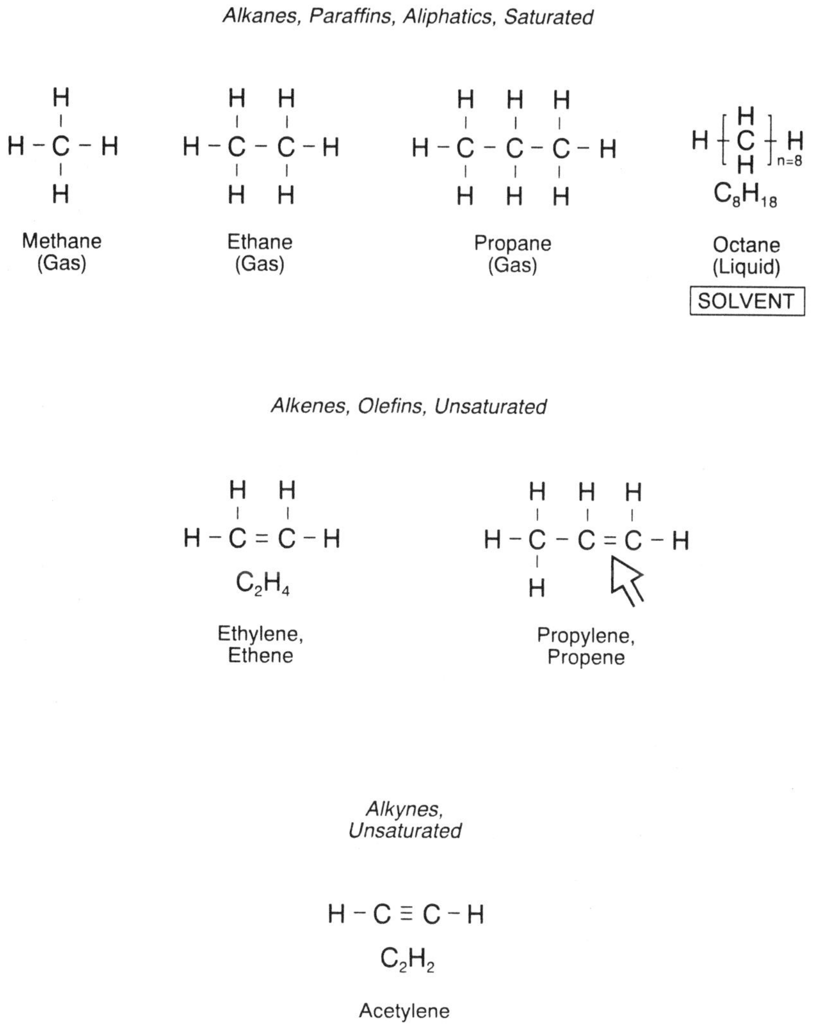

Alkanes, Paraffins, Aliphatics, Saturated

$$
\begin{array}{c}
H \\
| \\
H - C - H \\
| \\
H
\end{array}
\qquad
\begin{array}{c}
H \quad H \\
| \quad | \\
H - C - C - H \\
| \quad | \\
H \quad H
\end{array}
\qquad
\begin{array}{c}
H \quad H \quad H \\
| \quad | \quad | \\
H - C - C - C - H \\
| \quad | \quad | \\
H \quad H \quad H
\end{array}
\qquad
H \left[\begin{array}{c} H \\ C \\ H \end{array} \right]_{n=8} H
$$

Methane (Gas)   Ethane (Gas)   Propane (Gas)   Octane (Liquid)

C_8H_{18}

SOLVENT

Alkenes, Olefins, Unsaturated

$$
\begin{array}{c}
H \quad H \\
| \quad | \\
H - C = C - H
\end{array}
\qquad
\begin{array}{c}
H \quad H \quad H \\
| \quad | \quad | \\
H - C - C = C - H \\
| \\
H
\end{array}
$$

C_2H_4

Ethylene, Ethene

Propylene, Propene

Alkynes, Unsaturated

$$ H - C \equiv C - H $$

C_2H_2

Acetylene

Fig. 3.1 Hydrocarbon structures.

For each carbon atom in the main chain, there are two hydrogen atoms at the sides, and then two additional hydrogens at the ends. Thus a generalized formula for simple compounds can be written C_nH_{2n+2}. This whole class of straight-chain, simple *hydrocarbons* (hydrogen plus carbon) can be called either "alkanes," or "paraffins," or can be described as "saturated." (Outside of the field of organic chemistry, the word "paraffin" means a type of wax obtained from petroleum, but the organic chemical meaning is the same as "alkane.") It is useful to recognize these words, or at least know where to look them up, because they often appear in patent examples or other recipes. Another descriptive word on the diagram is "solvent," written inside a rectangular block on the figure. This block, whenever it appears on the diagrams, points out a typical use of that type of compound in ceramic technology.

The next classification of compounds in Fig. 3.1 is the "alkenes," which can also be described by the other words shown in quotes. The syllable at the end of each name, "-ene," means that there is a double bond between two carbon atoms, somewhere in the molecule. The open arrow points to whatever particular feature gives a certain group of compounds distinctive names. A generalized formula for these compounds can be written C_nH_{2n}.

Note that the structures at the bottom of the figure, which are sometimes called "unsaturated," have less hydrogen than the "saturated" structures at the top. The ones at the top have the maximum amounts of hydrogen possible, and that is why they are referred to as "saturated."

Alkyls. Fig. 3.2 shows a few structures of hydrocarbons, both saturated and unsaturated, in which some of the hydrogen has been removed from the molecule, leaving a *radical* . (Note that this is not a "free radical." That would have an "unsatisfied bond," with a single electron, but this does not.) The simplest member of the series is *methyl,* and the whole series is referred to as *the alkyls.* A radical of this type is sometimes abbreviated R, as described earlier. (Note again that, instead of being called a radical, it is sometimes called a *group.*)

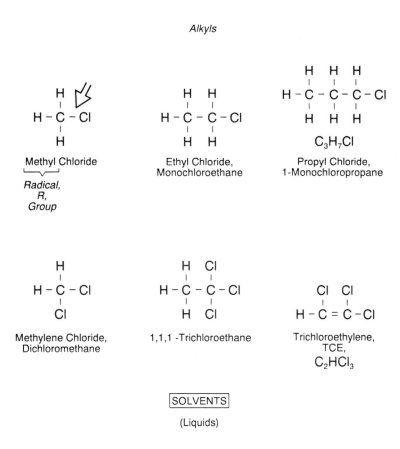

Fig. 3.2 Substituted hydrocarbons.

In the structures shown, one or more chlorine atoms have been "substituted" in place of the hydrogens, so these materials can also be called *substituted hydrocarbons*. Many other types of atoms could have been used instead of the hydrogen, examples being bromine, or other whole organic radicals, such as methyl or ethyl.

Mono- means that a single chorine is present, *di-* means two chlorines, and *tri-* means three. (See the lower drawings for examples.) Some other commonly occurring prefixes are *tetra-* four, *penta-* five, *hexa-* six, *hepta-* seven, *octa-* eight, *nona-* nine, and *deca-* ten. These terms are derived from the ancient Latin or Greek words for those numbers. The same prefixes also apply to the number of carbon atoms in the *backbone chain* of the structure, as in the example "octane" in Fig. 3.1. "Pentane" has five carbons, "hexane" six, etc. Among the larger carbon-carbon chains, various "C_{18} compounds" are important in ceramic processing. For example, the compound $C_{18}H_{38}$ is called "octadecane" and is a solid wax, sometimes used as a lubricant and as a binder. (The observant reader should note that this formula is another example of the C_nH_{2n+2} general rule written above.) Some of the smaller structures have special names of their own which are not derived from numbers, examples being "methane," "ethane," "propane," and "butane," as shown in Figs. 2.4 and 3.1. In many cases these names were in common usage during the early days of chemistry or alchemy, before the exact number of carbon atoms had been determined.

In Fig. 3.2, looking at the top three compounds, if the chlorine had been drawn at the left-hand end instead of toward the right, the materials would be exactly the same and would have the same names. This is because the actual molecules, at temperatures where they are liquids, are constantly rotating due to thermal motion. At some moment the chlorine could be at the left end, at some other moment at the top or bottom, etc. Even in the solid state, if frozen into crystalline form, the whole mass could be rotated by an experimenter, and the chemical nature of the material would not have changed.

The *systematic name* of the "1-monochloropropane" compound shown in the diagram could be thought of as "3-monochloropropane," the way it is drawn in the figure, since the number 3 indicates which carbon the chlorine is attached to, and the left-hand carbon in drawings is often called number one. However, the material shown in the drawing is the same as 1-monochloropropane, rotated so the chlorine is at the other end. One of the rules of the systematic naming system is that the numbers should be minimized, so if there is a choice of 1 or 3, the 1 would be chosen. For that reason the name of the compound shown must use "1," although the drawing can be made in several ways and usually has the chlorine to the right. (As pointed out earlier, please note that in this book, the structures and names are often purposely printed in several different ways, to accustom the reader to the several conventions found in other literature.)

A very important point is that a chlorine atom placed on the middle carbon of the propyl chloride would be a truly different compound, and no mode of rotation or other motion could transform the end position to the middle one. If the chlorine were actually on the middle carbon, the systematic name would be "2-monochloropropane" (not shown in the diagrams). More than just a single substituent (chlorine or other materials) can be present on the molecule, and three examples are shown on the lower line. The two chlorines in "methylene chloride" are equivalent and could have been drawn at several different places without changing it. It should be apparent to the reader that the structures in this figure are all simplified by "flattening," and that methylene chloride really has a tetrahedral structure which, in the liquid or gaseous states would be constantly rotating, bending, and vibrating.

Although there is a compound whose common name is "methylene chloride," there is no stable material called "methylene" with a carbon-carbon double bond, because there is only one carbon present. This is another example of how the common names are not completely systematic, where names include species that don't exist by themselves. (However, it should be remembered at this point that there is a stable "ethylene," as illustrated in Fig. 3.1.)

All of the hydrogens can be substituted by other elements such as chlorine, etc., as in the case of a compound (not shown) called "carbon tetrachloride" with four chlorines. If more than one carbon is in the "backbone chain," there is the possibility that a carbon-carbon double bond is present, as in the solvent whose common name is "TCE." Two of these "trichloro-" compounds are used as solvents in ceramic processing. These materials are not flammable, and therefore their large scale use was at one time considered preferable to that of the highly flammable hydrocarbons. They are mentioned in many patents and other publications, including several by the author. There are now some claims that they might cause environmental problems, so they are slowly being phased out of mass production use, in favor of other materials.

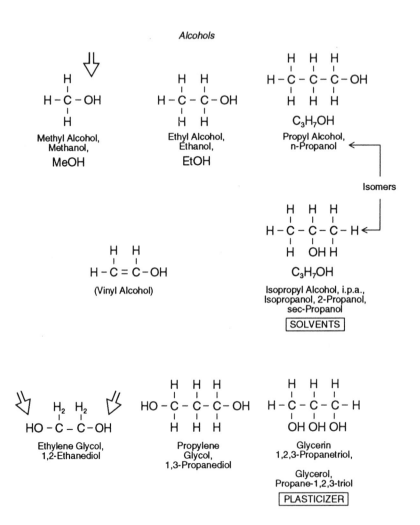

Fig. 3.3 Alcohols.

Alcohols. Figure 3.3 shows a series of compounds called "alcohols," whose names come from the fact that they contain OH groups (see the open arrow on the diagram). There are many different types, with names similar to the alkyls, but followed by the separate word "alcohol." Two common abbreviations are "MeOH" and "EtOH," the latter being "ethyl alcohol," which is the active ingredient of beer and wine.

"Vinyl alcohol" is drawn in parenthesis because it does not exist long enough to put in a bottle and store for future use. It can exist for a fraction of a second, so it is generated by one chemical reaction and then immediately used in another reaction. The final product usually has "...vinyl..." in its name, and such materials are often used in ceramic processing, an example being polyvinyl alcohol, which will be discussed in later sections.

Propanol, with three carbons, can have the OH ("hydroxyl") group at the end, in which case it is usually called "n-propanol." The "n" stands for "normal," but it is usually abbreviated. If the OH group is on the middle carbon, this is a different material, with a different boiling point, viscosity, etc. That is the more common form, because it is easier to manufacture. Note that both n-propanol and 2-propanol both have the same overall formulas, C_3H_7OH. They are called "isomers" of each other, with the prefix "iso-" meaning "similar" in ancient Greek. The 2-propanol variety has its OH group attached to a carbon which is itself attached to two other carbons, and therefore this alcohol is referred to as *secondary-,* abbreviated *sec -* or simply *s-* (often italicized). The "normal" variety of propanol could be called *primary,* but that word is only rarely used.

An example of a long-chain alcohol is "octadecanol," $C_{18}H_{35}OH$. If there is a carbon-carbon double bond somewhere in the chain, the material is "octadece*n*ol," with the "-en-" syllable used to indicate the unsaturated nature of the compound. These compounds are not shown in Fig. 3.3, but they will be discussed later, in connection with dispersants and injection molding.

Going back to a simpler structure for a moment, if the hydrogen is taken off the OH of ethanol, the remaining part is called an "ethoxy group," C_2H_5O-. The ethoxy could be combined with a metal such as sodium, forming a compound called sodium ethoxide. (This is a specific case of the more general term "alkoxide," which is an alkane combined with oxygen.) The element silicon is tetravalent, and four ethoxy groups can combine with it to form a compound called "tetraethoxysilane." This name is commonly abbreviated "TEOS," and the formula is $(C_2H_5O)_4Si$. (Note that "silane" is SiH_4, which is analogous to methane.) TEOS is well known in advanced ceramic technology, because of its use in "sol-gel" processes. It is also used a great deal in the semiconductor industry, as a starting material for the important process of depositing thin films of silicon dioxide by means of "chemical vapor deposition," usually known as CVD. (Another, equally valid name for TEOS is tetraethylorthosilicate, which can be understood better by referring to the structure of orthosilicic acid in Fig. 8.5.)

If a hydrogen atom is taken off the compound "butane" shown in Fig. 2.4, the resulting *group* or *radical* is called butyl, similar to ethyl or propyl. An oxygen added makes it butoxy, and this is one of the alkoxides. Now consider the element aluminum, which is trivalent and thus could combine with three alkoxide groups. An example of such a "tri-alkoxide" which is used to manufacture some new types of advanced sol-gel ceramics is "aluminum tri-sec-butoxide," $Al(OC_4H_9)_3$. This is abbreviated "ATSB," or sometimes just "ASB." The butyl groups are straight chains but have the oxygen bonded to the second carbon, and thus they are secondary. The compound could also be called "tri(sec-butoxy) aluminum," but that name is not ordinarily used. In the formula, the Al could be written either before or after the butoxy groups.

Aluminum can also combine with the butyl group directly, with no oxygen, to form $Al(C_4H_9)_3$. This is a "tri-alkyl," and it is most useful when the butyl group is the branched structure referred to as "isobutyl." The resulting compound is called "tri-isobutyl aluminum," abbreviated "TIBA," often used in the "chemical vapor deposition" of aluminum metal films on the surfaces of glass mirrors.[*]

---

[*] NOTE: It is hoped that this is not confusing to the reader. Many of these awkward names and acronyms do appear in the current literature. Ceramists will probably see such terms as TEOS, ASB, and TIBA cited more and more in the near future, as the costs of sol-gel and other advanced processes become lower, and as the new high performance materials that can be made from these chemicals become more competitive. (*Sol-gel* technology is now a whole subject by itself.)

More than one OH group can be attached to the main carbon chain (Fig. 3.3). If two are attached, the material is called a "glycol," and there is a series of these names, depending on the number of carbons. The "-ene" ending in the common name is not completely logical or systematic. It comes from the fact that ethylene glycol is often manufactured from ethylene, even though the double bond is no longer present in the final product.

Three OH groups on a three carbon chain is called either "glycerin," or "glycerine" with a final "-e," or "glycerol," which are all the same thing. Also, the systematic name can be either of the two shown in the figure (containing the numbers "1,2,3"). This compound is a useful *plasticizer,* particularly in making dry pressed ceramic bodies.

Carbonyl Compounds. In some compounds the oxygen is double-bonded to a carbon, and the resulting C=O group is called a *carbonyl* group. If this is in between two other groups such as methyl and ethyl (see example at the upper right corner of Fig. 3.4), the general type of

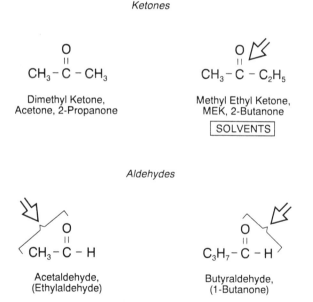

Ketones

$$CH_3-\overset{\overset{\textstyle O}{\|}}{C}-CH_3$$

Dimethyl Ketone,
Acetone, 2-Propanone

$$CH_3-\overset{\overset{\textstyle O}{\|}}{C}-C_2H_5$$

Methyl Ethyl Ketone,
MEK, 2-Butanone

SOLVENTS

Aldehydes

$$CH_3-\overset{\overset{\textstyle O}{\|}}{C}-H$$

Acetaldehyde,
(Ethylaldehyde)

$$C_3H_7-\overset{\overset{\textstyle O}{\|}}{C}-H$$

Butyraldehyde,
(1-Butanone)

Fig.3.4 Carbonyl compounds. The names in parenthesis are not ordinarily used.

compound is called a *ketone,* and that word usually appears in the common name, such as "methyl ethyl ketone." This material is used often as a solvent in ceramic processing and is sometimes abbreviated "MEK."

The suffix "-one" (pronounced like the word "own") in the name of the overall compound is associated with the C=O group, and the open arrow points to this. The systematic name for MEK is "2-butanone," since the carbon-carbon backbone is that of butane, but there is a double-bonded oxygen at the second position. It is never called "3-butanone," since organic names use the smallest possible numbers.

If both radicals are methyls, the compound is "dimethyl ketone," as at the upper left corner of the figure. This particular material is usually known by its common name, "acetone," and only rarely by its systematic name. The systematic name of acetone is 2-propanone, since it is propane with a double-bonded oxygen at the second carbon position.

(It should be noted by the reader that, while the oxygen of an alcohol has two bonds, only a single bond goes to carbon, so it its not a ketone. The other bond goes to hydrogen.)

If a four-carbon chain has a C=O at the end, rather than being between two other radicals, it is referred to as an "aldehyde." A diagram of the structure is shown at the bottom right corner of Fig. 3.4. The systematic name of this material is "1-butanone," but it is almost always called by the common name "butyraldehyde." Its chemical behavior is quite different from that of ketones, because there is a hydrogen atom attached to the C=O carbon, and hence the different type of name. Aldehydes are not used directly in ceramics, although "polyvinyl butyral," which we do use in ceramics, is made from an aldehyde. This will be discussed in a later chapter on binders.

The aldehyde with two carbons is called "acetaldehyde." In both this compound and in acetone, the prefix "acet-" means the somewhat complex CH_3CO- group[**] This same prefix will appear again in the next section, in the common name for *acetic acid,* which also contains the CH_3CO- group (see open arrow at upper left hand of Fig. 3.5).

[**] The double bond of the C=O is not shown in this method of writing the formula, but it is assumed that the reader knows that it is there. Organic chemists often use formulas which assume considerable knowledge on the part of the reader.

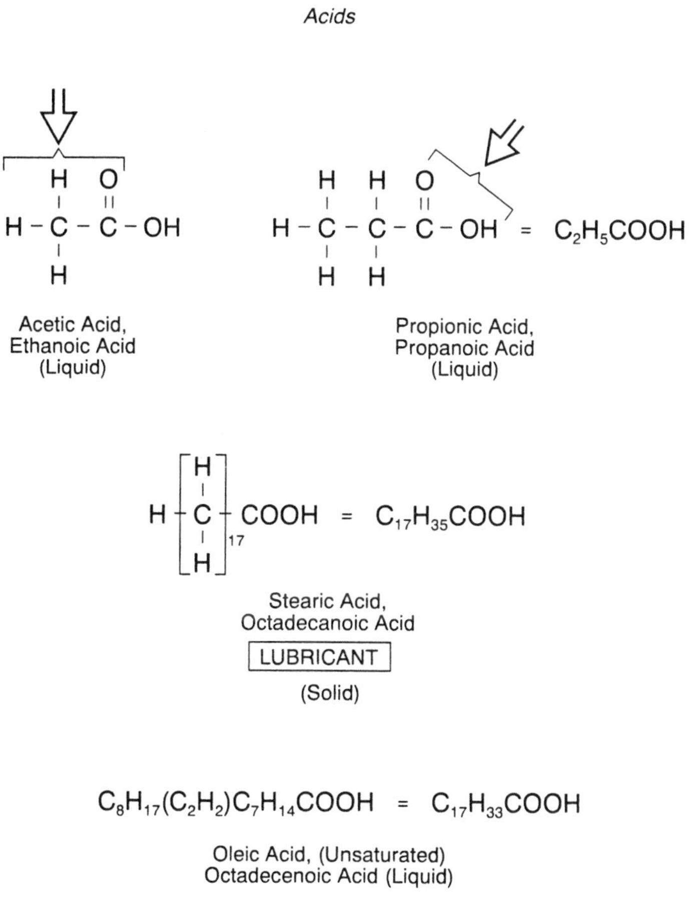

Acids

H   O
| ||
H – C – C – OH
|
H

Acetic Acid,
Ethanoic Acid
(Liquid)

H   H   O
| | ||
H – C – C – C – OH   = C_2H_5COOH
| |
H   H

Propionic Acid,
Propanoic Acid
(Liquid)

$$H \left[\begin{matrix} H \\ | \\ C \\ | \\ H \end{matrix} \right]_{17} COOH \quad = \quad C_{17}H_{35}COOH$$

Stearic Acid,
Octadecanoic Acid

| LUBRICANT |

(Solid)

$$C_8H_{17}(C_2H_2)C_7H_{14}COOH \quad = \quad C_{17}H_{33}COOH$$

Oleic Acid, (Unsaturated)
Octadecenoic Acid (Liquid)

Fig. 3.5 Carboxylic acids.

Carboxylic Acids. If a carbonyl is attached to an OH, as in the upper right hand corner of Fig. 3.5, the resulting –COOH is called a *carboxylic acid* group. This simplified method of writing the formula as-
sumes that the reader knows the true structure, in which the first oxygen has a double bond to the carbon. If the whole compound is dis-

solved in water, the hydrogen comes off easily, producing mobile hydrogen ions and *Bronsted-Lowry acidic* behavior. Most of these materials are only weak acids compared to the better known sulfuric and hydrochloric acids, especially as the molecules get bigger.

The total number of carbons is counted to determine the name, and the carbon inside the $-COOH$ group is included in the count. Thus, "octadecanoic acid" has 17 carbons in the main chain, plus one more in the carboxylic acid group, so it is sometimes referred to as "a C_{18} acid." The common name, stearic acid, is derived from the ancient Latin word for cow, and in fact stearic acid is a major ingredient of the fat in cows and other animals. (This word is not to be confused with "steric" without the "a," which is derived from an ancient Greek word for 3-dimensional. "Steric hindrance" is a phrase that will be used later in this book.) Stearic acid is an important lubricant for extruding and dry pressing ceramic bodies.

"Oleic acid" is another C_{18} structure, with a carbon-carbon double bond in the middle. Sometimes the formula is written with a $C=C$ to indicate the double bond, although it is not always shown. The systematic name has the syllable "-en-" (pronounced "een") instead of "-an-" in order to indicate the "alkene" double bond. With most of the organic compounds in this general range of chain lengths, the unsaturated compound is a liquid, while the saturated counterpart is a solid. The natural fats containing unsaturated carboxylic acids are usually found in fish and plants, while the saturated varieties are usually in animals. It has recently become apparent that the unsaturated fats are healthier foods for humans, in terms of both cancer and heart trouble, so the words "grams of unsaturated fat" now appear on the labels of many food containers.

Cyclohexane and Related Structures. Figure 3.6 shows some organic compounds in which the chains of carbons loop around to themselves again, making rings. Regarding the bond angles, it can be seen from the figure that a simple, flat hexagon contains six angles, each of which is 120 degrees. However, the natural angles between the four unstrained bonds that attach to a carbon atom are each 109 degrees, as was shown previously in Fig. 2.4. Therefore, the six-carbon molecule called "cyclohexane" in Fig. 3.6 requires some twisting and bending of the carbon-carbon bonds in order to complete the structure. (Actually,

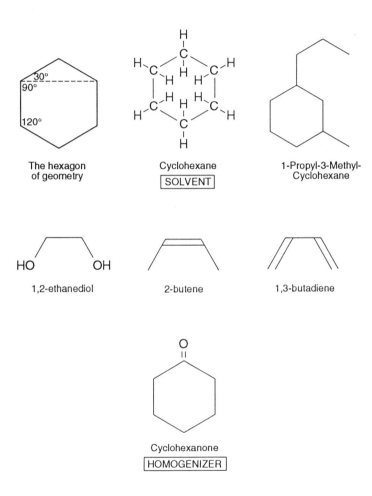

The hexagon
of geometry

Cyclohexane
SOLVENT

1-Propyl-3-Methyl-
Cyclohexane

HO   OH

1,2-ethanediol

2-butene

1,3-butadiene

Cyclohexanone
HOMOGENIZER

Fig. 3.6 Cyclic compounds and some abbreviated structures.

the real molecule is not perfectly flat, although the simplified diagram in Fig. 3.6 has been flattened for the sake of simplicity.) Ring compounds containing three carbons, four carbons, and other numbers of carbons also exist, but the amounts of bond-twisting and bond-bending that are required geometrically to complete the rings are greater than with six carbons. Therefore, the six-carbon structure occurs commonly in nature and is also easier to make synthetically. Cyclohexane is inexpensive and is therefore used as a solvent in some ceramic manufacturing processes.

Chemists have to draw so many structures that they tend to adopt a shorthand drawing system in which the hydrogen atoms are left out, and sometimes even the carbon atoms are not drawn. The figure shows a simple hexagonal compound in which only the most important bonds are drawn in, as in the case of the compound 1-propyl-3-methylcyclohexane.‡ Not only does the chemist not bother to draw the hydrogen atoms, but the bonds to the hydrogen atoms are also not drawn, and sometimes only the carbon-carbon bonds actually show up in this shorthand type of drawing. While regular cyclohexane has twelve hydrogen atoms bonded to it, the 1-propyl-3-methylcyclohexane molecule has only ten hydrogen atoms (not shown) bonded to the central 6-carbon ring, because two of the original twelve have been *substituted* by propyl and methyl radicals. Similarly, each of the radicals has one fewer hydrogens than propane or methane would have. As far as the numbering is concerned, the most important "substituent," propyl or methyl, gets to be number one; in this case, propyl is the largest, so it gets the honor.

The *substituent* propyl radical is drawn in zig-zag fashion, just as the backbone structure actually is (see butane in Fig. 2.4). The number of carbon atoms can be quickly counted by observing the number of 109° bends in the backbone structure (four carbons in this case, of course). However, it is not always true that just the carbon-carbon bonds are shown in shorthand drawings. Generally, whatever bonds must be shown in order for the reader to quickly interpret the composition will be drawn in. An example is ethylene glycol (1,2-ethanediol),

---

‡ In the modern systematic name, there is no space between the methyl and the cyclohexane. However, in the older common names, there can be a space. In Fig. 3.6, the hyphen there is only because the word is too long for one line.

in which the carbon-oxygen bonds are also shown, since these are necessary in order to indicate the presence of the carbons at the ends of the backbone. (No attempt is made to make accurate 109° angles in the drawing, since this is just a 2-dimensional projection of what is really a 3-dimensional structure.) Note that the oxygen hydrogen bonds are not shown, because the OH group is so common that it is often drawn as a single entity. In reality, the two bonds of the oxygen, one to the hydrogen atom and one to the nearest carbon, are at approximately 90° angles to each other, but the chemist usually does not bother to show this accurately, because it has little significance to organic chemistry in most cases.

The zig-zag, highly simplified diagrams do not have to be associated with ring structures. For example, 2-butene is often shown in modern organic chemistry publications as the simple line drawing shown in Fig. 3.6. It should be noted once again that the lowest possible number is chosen for the name (2 instead of 3 in this case). Another example is in the naming of butadiene, a commercially important chemical which is used in huge quantities to make rubber and other polymers to be discussed in a later section. This compound can only exist in one *isomeric* form, the 1,3 form illustrated in the figure, because the 1,2 form (not shown) with a pair of immediately adjacent double bonds is not stable and would immediately decompose. Although this fact might not be obvious to the average reader, it is well known to organic chemists. Therefore, in the interests of efficiency and concise writing, the "1,3" positioning numbers for the double bonds, are not ordinarily used by chemists in this case, and just the name "butadiene" without the numbers is sufficient to uniquely describe the compound.

If one of the carbon atoms has oxygen double-bonded directly to it, the compound is a *ketone,* similar to the materials in Fig. 3.4. The cyclic ketone at the bottom of Fig. 3.6 is *cyclohexanone,* which is sometimes used in ceramic processing as a special type of plasticizer called a "homogenizer," and this type of additive will be discussed later in the section on tape casting.

When there are two substituents on ring compounds, the relative positions must be designated in the name, and there are several ways to do this, as shown in Fig. 3.7. Three of the most common ways are as follows:

Alpha (a) means that the substituent (chlorine, bromine, methyl, etc.) is bonded one carbon away from the carbon to which the important item (oxygen in this case) is bonded; *beta* is similar but two carbons away, and *gamma* is three carbons away;

*Ortho** means that the two substituents (chlorine and oxygen in this case) are bonded one carbon away from each other, *meta* means two carbons away, and *para* is on the opposite carbon;†

2, 3, and 4, are the systematic designations, in which the oxygen is part of the backbone structure and therefore does not have to be given its own number, but 2-chlorocyclohexan-1-one is also a possible name for the alpha material.

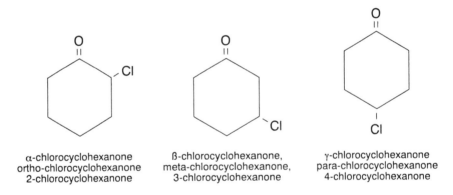

α-chlorocyclohexanone	β-chlorocyclohexanone,	γ-chlorocyclohexanone
ortho-chlorocyclohexanone	meta-chlorocyclohexanone,	para-chlorocyclohexanone
2-chlorocyclohexanone	3-chlorocyclohexanone	4-chlorocyclohexanone

Fig. 3.7 Names of ring structures having two substituents. Note: ortho, meta, and para are often abbreviated as *o, m,* and *p.*

---

* In many scientific articles, prefixes such as *alpha, ortho,* and also *sec* and *tert* are italicized.

† Ortho, meta and para came from ancient Greek, with the approximate meanings of standard, mid-, and parallel.

Other Positional Terms. Figure 3.8 shows some special names that are not obvious. When four or more carbons are strung together, one of the ways they can be configured is isobutane, as shown in the upper left diagram. There is a systematic name for it, 2-methylpropane, although there could be no confusion as to the structure, because that is the only possible "branched" (iso-) configuration. The material that was shown in Fig. 2.4 is properly called "normal butane" or more often "n-butane," although just plain "butane" is also used. However, a tank of butane is likely to contain both isomers, unless it is

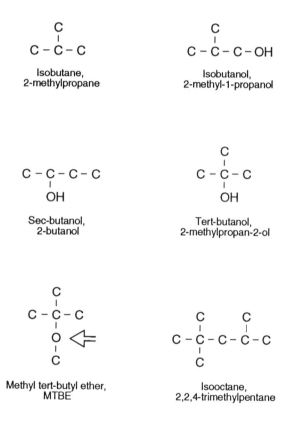

Fig. 3.8 Some special positioning names and an ether.

labeled specifically as being only the normal butane or the only isobutane type. Occasionally the tank will be labeled with the plural form "butanes," if it contains both isomers. Isobutan*ol* is named after its isobutane parent compound, as opposed to using that name for *sec*-butanol. It can also be called isobutyl alcohol.

The "tertiary" isomer of the alcohol derives its name from the fact that three carbons are attached to the one having an important group (OH) on it, as compared to "secondary," and it is usually abbreviated *tert-* or *t-*, sometimes italicized. Tertiary compounds often are found to have unusual properties because of *steric hindrance,* in which other large groups of atoms would have difficulty approaching the OH group, for example to make hydrogen bonds. Therefore, the boiling point and chemical reactivity are quite different from those of the normal or the secondary isomers. A compound which might become important in the future is MTBE, which is now being added to lead-free, low-pollution gasoline instead of the older tetraethyl lead antiknock compound. When it burns, the highly *branched* structure tends to break up into somewhat stable *free radicals* such as t-butyl. These combine quickly with the free radicals that are involved in excessively fast explosions ("knocking" or "pinging") in automobile engines when the spark is advanced too much or the cylinder walls are too hot. The destructive type of explosion is thus slowed down to the desired slower type, transferring its energy into piston motion instead of only heating the top of the piston. Also, MTBE aids in burning up the last bits of gasoline, even though it slows down the reaction somewhat. For these reasons, MTBE is being recommended as a replacement antiknock additive instead of tetraethyl lead. The reader should note the word "ether" in the name, which refers to the oxygen linkage between two carbons as indicated by the arrow.

Another branched chain is found in isooctane (pronounced with three syllables, "iso-oc-tane"), which is also an antiknock ingredient of gasoline, for the same reason: the highly branched structure. The octane molecule can exist in other isomers, including n-octane with an unbranched chain. The "octane rating" of gasoline is based on 100% isooctane being "100 octane." Ordinary gasoline contains several of these isomers, in addition to heptane, nonane, etc., and it has less than a 100 octane rating. The concept of chain *branching* is beginning to find application in ceramic processing, particularly in the optimization of dispersants, and *free radicals* are involved in binder burnout, so these are not the last times the reader will see these terms and also *steric hindrance* being used.

Figure 3.9 shows still another series of names that occasionally are encountered by users of organic chemicals and probably will be seen more and more in the future by the general public. Symmetrical, or *sym-* (but never just *s-*, since that means secondary) is a designation whose meaning is apparent when the structure is shown. It is only used in a few cases where several atoms of a substituent (fluorine in

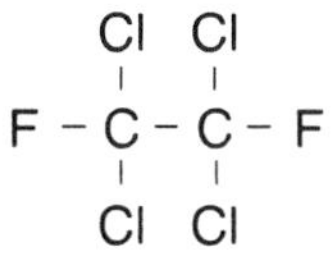

$$
\begin{array}{c c}
\text{Cl} & \text{Cl} \\
| & | \\
\text{F} - \text{C} - \text{C} - \text{F} \\
| & | \\
\text{Cl} & \text{Cl}
\end{array}
$$

Sym-diflurotetrachloroethane,
1,2-difluoro-1,1,2,2-tetrachloroethane

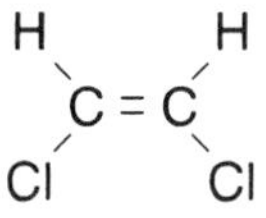

$$
\begin{array}{c c}
\text{H} & \text{H} \\
\diagdown & \diagup \\
\text{C} & = \text{C} \\
\diagup & \diagdown \\
\text{Cl} & \text{Cl}
\end{array}
$$

Cis-1,2-dichloroethene

$$
\begin{array}{c c}
\text{Cl} & \text{H} \\
\diagdown & \diagup \\
\text{C} & = \text{C} \\
\diagup & \diagdown \\
\text{H} & \text{Cl}
\end{array}
$$

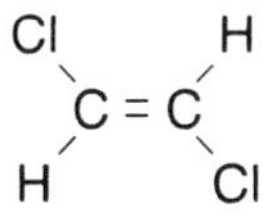

Trans-1,2-dichloroethene

Fig. 3.9 More special positioning names.

this case) might have been on the same carbon but actually are not. It is interesting to note that this compound is Freon 112, one of the famous materials which have been installed in large amounts in refrigerators, air conditioners, and industrial cleaning equipment, and which are slowly leaking out into the surrounding air. (Most of the other Freons are not symmetrical.)

Some Special Environmental Considerations. Environmental activists suspect that these "chlorofluorocarbons" (or "CFCs") are destroying the stratospheric ozone layer and causing the "ozone hole" that allows more ultraviolet sunlight to penetrate, thus presumably causing increased skin cancer among farmers and other people. The Governments of many industrialized countries, including the United States, have agreed to enforce the elimination of these chemicals by a stepwise timetable. This elimination process has begun, at great cost to industry and consumers.

Whether man-made CFCs are a significant threat to the ozone layer compared to purely natural volcanoes, sunspots, and ocean algae is quite debatable, as well as the question of whether a "hole" in the ozone layer is really a significant threat to humanity. For example, ocean algae have been reported to produce hundreds of times more organic chlorides and bromides than the world's human population does.[1,2,3] The recently discovered ozone hole over Antarctica and Chile is thought to be influenced to some extent by volcanoes such as Mt. Erebus and Mt. Pinatubo and also by an eleven year sunspot cycle. However, even if a contributory cause was also man-made CFCs, the wavelengths of ultraviolet light allowed through the ozone hole (<300 nm) are mainly not in the range (302 to 436 nm) that causes skin cancer in humans.[4,5] Unfortunately for the users of CFCs and other chlorine containing organic compounds such as 1,1,1-trichloroethane, whatever scientific data are considered contrary to the contentions of

---

[1] A. M. Wousmaa, et al., Science, 249 (1990) 160.

[2] G. W. Gribble, Chem Ind., 6 (1991) 591.

[3] G. W. Gribble, Chem. & Engrng. News, (Nov. 9, 1992) 3.

[4] R. Setlow, et al., Proc. Nat'nl. Acad. of Sci. (July 15, 1993) 6666.

[5] Anon., Environ. Sci. and Tech., 27 (1993) 1956. Also, there has been no increase in skin cancer in Chile under the ozone hole, in spite of alarmist-inspired rumors to the contrary: see C. Sims, N.Y. Times, March 3, 1995, page A4.

environmentalist extremists are largely being ignored by the popular news media and therefore by political decision makers.** The results of new studies telling us that "everything is OK, after all" rarely get as much attention as the original study that reported "danger, environmental pollution."

There is far more to be said on both sides of this debate, including possible effects of various "chlorocarbons" on sexual reproduction. Certainly the ceramic engineer must be careful not to allow chemical waste to escape into the environment if this will cause harm. However, distinguishing a true cause of harm from the background, quite random occurrence of such things as skin cancer is a scientifically difficult problem. Even the simple concepts of standard deviations and correlation coefficients are not usually well understood by news media headline writers and legislature politicians. In science-intensive fields such as this, the emotions of poorly informed people sometimes overwhelm logical debate. Therefore it is suggested that the reader take note of any future data published on this subject, and possibly take an activist approach on whatever side seems to have scientific merit. Leaving such decisions entirely up to people who are not scientifically educated or not practical minded is going to be an unwise policy for the future, especially as environmental issues overlap ceramic engineering more and more. This overlap is likely to become particularly important in the realm of chlorocarbon chemistry. More about this important subject is discussed at some length in the chapter on solvents.

Figure 3.9 also illustrates the *cis-* and *trans-* concepts in organic chemical nomenclature. (Note that *cis-* and *trans-* are often italicized. Also, it should be recalled from the labels in Fig. 3.1 that ethene is the same thing as ethylene.) If the two structures shown in the bottom part of Fig. 3.9 had single bonds and were therefore ethane "derivatives" instead of being derived from double bonded ethene, the two structures would be the same material, simply shown in different views. The left hand carbon and its substituent chlorine could sometimes rotate, without the right hand carbon and chlorine moving at all. One structure could then become the other, and there would not be two separate

---

** For example, politicians are mostly ignoring the fact that volcanoes such as Erebus generate huge amounts of HF (C&ENews, Feb. 13, 1995, p.5), and the hot gases go directly to 70,000 ft., above rain clouds (Sci. News, Oct. 8, 1994, p.228).

compounds for us to consider. However, the carbon-carbon double bond is unusual in that the ends can not rotate relative to each other, and therefore the *cis-* and *trans-* forms are really different materials, with different boiling points, chemical reactivity, etc.‡

Since oleic acid (Fig. 3.5) has a double bond near the middle of the molecule, a section of the carbon chain sticks out at an angle where each of the chlorines is in Fig. 3.9. Therefore there is a *cis-* oleic acid and also a *trans-*, which are quite different compounds. The unsaturated acids occurring in natural fish oils and vegetable oils are mostly *cis-*, and when consumed by people in food, these do not cause the liver to produce excessive cholesterol, so they are quite benign. By comparison, the ones in chemically treated margarine are *trans-* acids. These have recently been shown to trigger the production of cholesterol, to the extent that many people with a tendency toward congestive heart trouble from excessive cholesterol are now being told by medical doctors to avoid the use of margarine.[6] This is ironic, because margarine had previously been recommended by researchers as a harmless substitute for butter, because margarine contains esters of unsaturated acids (oleic, etc.) while natural butter contains esters of the saturated acids (stearic, etc.), and the latter were known to stimulate the body's production of cholesterol, while the former were *all* thought to actually decrease the harm from cholesterol. More recently obtained knowledge indicates that not all unsaturateds are the same, and the *trans-* type is slightly harmful to certain people.

As more is publicized about these sorts of distinctions, the average nontechnical person is becoming conversant with many of the subtleties of organic chemical nomenclature, a clear example being the appearance of the term *trans-* in popular news media such as reference 6. In fact, the life expectancy in industrialized countries is still continuing to rise, apparently at least partly because of the widespread utilization of such knowledge. A major challenge to educators is to help

---

‡ Another point for the reader to note carefully is that, if both chlorines were on the same carbon, as in the case of 1,1-dichloroethene, then the chlorines could be drawn at either the left or right end of the structure, since both would show the same material just being rotated around a vertical axis.

6 J. E. Brody, New York Times (March 24, 1993) C-13.

nontechnical voters distinguish scientific truth from pseudoscientific myth, and to use the new knowledge to steer in the right direction. So far, more benefit seems to be coming from the new awareness than harm, as evidenced by the upward life expectancy trend. However, as more synthetic compounds that never existed in nature become newly present in everyone's environment, with essentially no escape from exposure, the potential for a seriously harmful effect that takes decades to detect begins to take on perilous dimensions. For example, how many people in the industrialized world have never eaten margarine (at least in cooking), or never eaten chlorocarbon pesticides on fruit? This question, coupled with the fact that potential harm from such materials has only recently become evident,[*] is an understandable motivation for environmental activists to sometimes take rather extreme sounding positions.

Cyclic Compounds With Double Bonds. If the cyclohexane molecule shown in Fig. 3.6 had a carbon-carbon double bond put into it (and therefore two less hydrogens), it would be called cyclohex*ene* (not shown in Fig. 3.6). Then, if another double bond were formed, it would be cyclohexa*diene*, which is illustrated in Fig. 3.10. (Note that, just as in the case of the 1,3-butadiene structure shown in Fig. 3.6, the two double bonds can not be next to each other.) Also, a five-carbon ring called cyclo*pentane* can be given two double bonds, making it cyclopenta*diene* (Fig. 3.10), which finds considerable use in the technology of *chemical vapor deposition.,* or *CVD.* Ceramists are hearing about cyclopentadiene more and more, for the deposition of coatings to be used in electronics, fiber optics, cutting tools, rocket nozzles, etc. The compound cyclopentadiene makes special kinds of bonds with several metals. For example, the compound called tris(cyclopentadienyl)yttrium (not shown in the figure) can be easily formed and then boiled to make a vapor, and at a still higher temperature it can then be decomposed to deposit yttrium metal. If a small amount of oxygen is present, yttrium oxide can be deposited as a ceramic coating, which is used because of its transparency to microwaves.

---

[*] These two examples are only presented here because of their obviously widespread use. At present, there is no solid evidence that *small* amounts of these two types of material are at all harmful.

The prefix *tris* means three cyclopentadienyls, and the whole compound is sometimes referred to as a *complex*. The latter word, used as a noun, means a compound in which a chemical element (in this case yttrium) has one or more other materials (in this case cyclopentadienyls) bonded to it, usually by Lewis acid-base coordination in addition to more conventional bonding, and that is the situation here.[†] The

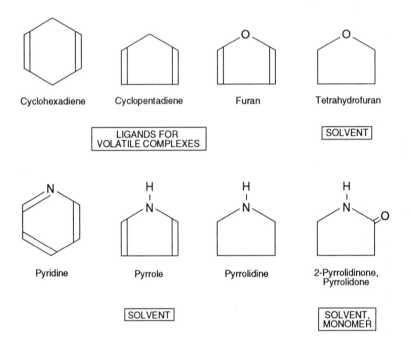

Fig. 3.10 Other ring structures. Note: when a methyl group is attached directly on the nitrogen of pyrrolidone, this is the solvent N-methylpyrrolidone, or NMP.

[†] Another *complex,* which perhaps is more familiar to the reader, is the ferrocyanide ion, in which the cyanide groups are attached to the iron by both conventional ionic bonds and also by Lewis acid-base coordinate bonds.

word *ligand* within the box in Fig. 3.10 means the material such as cyclopentadienyl that is bonded to the metal in the complex. Instead of *tris,* the prefix *bis* would mean two, and the prefix *tetrakis* means four. Examples are bis(cyclopentadienyl)ruthenium and tetrakis(cyclopentadienyl)titanium.

These materials and their names are possibly unfamiliar to many engineers who might not be keeping up to date on the newest developments in chemistry. However, chemistry-literate ceramists are searching the world for these unusual structures and for these rare chemical elements, in attempts to obtain the best possible properties. The world of technology has become extremely competitive, with the race being won by materials that are only slightly better than the rest. At the same time the world of technology has become extremely lucrative, with the winners sometimes gaining great recognition and wealth. The modern ceramist should not be afraid to join in this search for "sophisticated" technologies, meaning that the scientific principles involved are quite far from the ordinary, and that a great deal of knowledge is required to compete. These sophisticated technologies are actually available to all of us. Appendix III lists a few readily accessible corridors, which might help lead to the fast changing pool of knowledge that probably will be useful in making the ceramic-based materials of the future.

Figure 3.10 also shows a few other ring compounds which are finding use in polymer chemistry. Three of them are unusually powerful solvents for the kinds of polymers that might make good binders in ceramic processing. They are already used to some extent in inks for printing three-color photographs in books and magazines, and possibly they might become involved in ceramic decorating or in making ceramic-based electronic circuitry.

Benzene Ring Compounds. If a six-carbon ring has three double bonds and the appropriate hydrogens, it is called benzene, as shown at the top of Fig. 3.11. The two structures shown are really the same material. They could be gotten from each other by rotating the molecule around a vertical axis. However, what happens more often is that the electrons in the ring move, instead of the whole atoms moving. The real electronic structure changes back and forth between the two configurations shown. The fact that there are two equivalent structures which can interchange simply by the motion of electrons is called "resonance." Because of a sophisticated principle in the science of quantum mechanics (the "uncertainty principle"), there is no way to experimentally determine which structure is present at any given time. For this reason, chemists most often draw benzene as the hexagon illustrated with a circle inside it, showing that the electrons are at indeterminate positions. In fact, the single bond electrons and the double bond electrons are all merged together, in "pi" type, donut-shaped orbitals. The resonance phenomenon makes the *benzene ring* slightly

Aryls, Aromatic

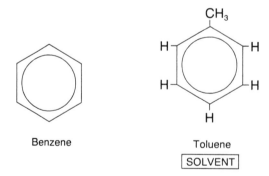

Resonance Interchanges

Benzene   Toluene
                            SOLVENT

Fig. 3.11 Benzene ring compounds.

more stable than it would be if one simply added up the bond energies of the nine carbon-carbon bonds.** Therefore, benzene rings appear in many facets of nature, and also they are often found in man-made chemical compounds. The rings are usually drawn by chemists without specifically indicating the carbon atoms or the hydrogens, although there is nothing wrong with drawing them in. These unsaturated, resonating ring structures are referred to as "aryl" compounds (see top of Fig. 3.11), just as the saturated forms would be called "alkyls." An older word, which is still used to a lesser extent for describing this family of materials, is "aromatic." Many of them have a distinctive aroma or odor, although it is not a particularly strong odor, and the word aromatic was given to them in the early days of chemistry, before the various attributes of odor were correctly understood. (It is now known that most of the strongest pleasant odors are due to the presence of aldehydes and esters in organic chemical compounds; esters will be described later in this chapter.)

** Through the science of quantum mechanics, we have learned that nature favors cases where electrons can occupy any of *several* different equivalent positions, and materials will tend to go into such states. Thus structures such as benzene are more stable than cyclohexane. There are many other examples of resonance stabilization in chemistry. For instance, the ammonium ion shown in Fig. 2.1 exhibits resonance between the four hydrogens, which all become equivalent and are therefore unusually stable. Another example is the ionized carboxylate group –COO⁻. The two oxygens become equivalent, and the negative charge is actually shared between them, not on just one of them. Chemists often draw the structure of this, not with a double bond to one oxygen and a single bond to the other (from the carbon), but instead with a single bond going to each oxygen and a dotted line also going in parallel to each oxygen. The dotted line represents a bond that is shared equally between the oxygens. This stabilizes the ion and makes –COOH ionize quite readily into a –COO⁻ structure. That is the reason why alcohol does not ionize as much as acetic acid: the latter has resonance possibilities, while the former does not. This is also the reason why sulfuric acid ionizes readily: the negative charges are actually shared among several oxygens, stabilizing the ionized structure, and thus making ionization more likely. In fact, it is also part of the reason why a crystal is more stable than an amorphous glass, and why electrons can travel so easily between metal atoms in highly perfect crystalline structures: the electrons are shared equivalently between the many metal atoms. However, imperfections such as vacancies tend to spoil the equivalence, making the electrical conductivity become poorer.

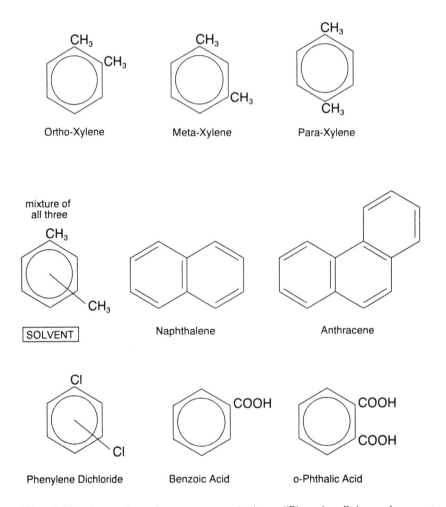

Fig. 3.12 Aromatics of greater complexity. "Phenylene" is analogous to "methylene" in Fig. 3.2.

Pure benzene itself is known to be a "carcinogen" (cancer-causing agent), particularly as a cause of leukemia and liver cancer among some of the factory workers who were exposed to it in large quantities before the danger was known. (The author was one of those people heavily exposed, during two summer jobs he held while a student about 40 years ago. Blood analysis showed the presence of benzene many months later, along with an alarmingly low white blood cell count, but no known permanent harm was done in this particular case.) Because of the toxicity, benzene itself is hardly ever used by modern ceramists. However, some less toxic compounds that have the benzene ring as a component are used commonly in ceramics as well as in many other technologies.

The addition of a methyl group to the benzene ring yields the compound *toluene,* which is often used as a solvent in ceramic tape casting. Factory workers exposed to toluene have not shown higher than usual incidences of cancer, possibly because the human body can metabolize toluene into carbon dioxide and water, and the material therefore does not remain in the bloodstream as long as benzene does. Some evidence indicates that bacteria normally in the digestive tract can pull off the methyl group, and the ring itself then opens up and is metabolized. The same thing is true of the three forms of *xylene* shown in Fig. 3.12, and they do not remain in the human body for long times. Therefore toluene and xylene are now used in large quantities as solvents, in ceramic processing as well as other fields. More discussion about the potential toxicity of these materials will be presented in the chapter on solvents.

Toluene, xylene, and other benzene ring compounds can be drawn either with the circles inside, or by showing the individual bonds, although the circles are preferred. Xylene can exist as three different isomers, each having slightly different boiling points, etc. Since these differences are slight, the isomers are not usually separated, and a commercially available drum of xylene usually will contain all three. This fact is sometimes indicated for chemists by having the methyl group appear to be bonded to the center of the ring, as illustrated in Fig. 3.12. It is assumed that the reader of such a diagram knows the chemical shorthand and realizes that the methyl groups are really bonded to the outer edges of the rings, at either of three *o-, m-,* or *p-* positions. This mixture of xylenes is used increasingly for manufacturing ceramics in the electronics industry, for reasons that will be explained in the sections on tape casting.

(Readers who are ceramists might take note of the fact that none of the benzyl or phenyl compounds named in the following paragraph are used much in ceramic processing. These names are only presented here for background knowledge.)

A benzene ring with an OH group on it is not considered to be an alcohol, because it has special properties. For example, it is a weak Bronsted acid, like stearic acid is, while most alcohols are neutral. This compound, C_6H_5OH, is called phenol, and an older name is carbolic acid. It is often used by dentists to sterilize a freshly drilled cavity before putting in a filling, and its distinctive odor would probably be recognized by many readers as "the dentist's office smell." If the OH of phenol is replaced by chloride, the compound C_6H_5Cl is called phenyl chloride, instead of benzyl chloride. However, it can also be called chlorobenzene, even if that is somewhat confusing. Because early chemists were not always sure of the structures of new compounds that they made, the name benzyl chloride was given to C_6H_5-CH_2-Cl (usually written without the dashes that have been drawn in here for clarity). Other common names for this compound are tolyl chloride and chlorotoluene. The compound C_6H_5-CH_2-OH is called benzyl alcohol. In other words, "benzyl–" is C_6H_5-CH_2- but "phenyl–" is C_6H_5-, and "–benzene" is also C_6H_5- when there is something like chlorine attached to it, like in chlorobenzene.

Two benzene rings can share a common side, making *naphthalene,* as shown in the diagram. This and *anthracene* have the physical property of "subliming," in which the heated solid can evaporate directly to the gas phase without first going through melting to the liquid state. Naphthalene was used for a while as a binder for extrusion and injection molding of ceramics having shapes that are difficult to produce, because the binder could be removed directly, without first melting and therefore "slumping" (deforming under the influence of gravity) while liquid. This is the same reason that naphthalene is used to make mothballs — its property of subliming. An even more common material that sublimes is frozen water in the form of snow, and meteorologists say that we are lucky that most snow actually sublimes in the springtime, rather than melting, because severe flooding would be common if all of it had to liquefy before eventually evaporating. Naphthalene is now suspected of being toxic to the human liver, although not necessarily as a true carcinogen. For this reason it is no longer used in large scale ceramics manufacturing, although it can still be legally sold for use as a moth repellent, where the human exposure would only be at low levels.

The benzene ring can be part of various organic acids, and two are shown in the figure. The *para-* form of phthalic acid (Fig. 3.20) has a special common name, "terephthalic acid," and this is used to make polyester film for ceramic tape casting, to be explained further in later sections of this book.

Nitrogen Compounds. If the OH were taken out of ethyl alcohol and replaced by a nitrogen and its appropriate hydrogens (two, in this case), the new compound would be *ethylamine,* as shown in Fig. 3.13.

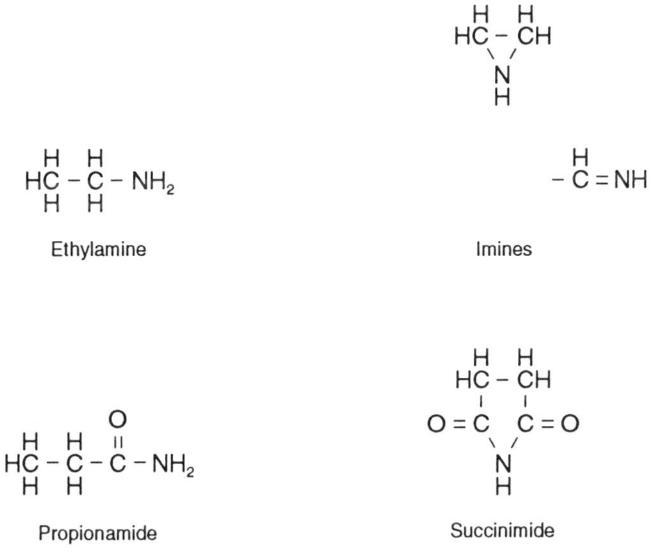

$$
\begin{array}{cc}
& \begin{array}{c} H \quad H \\ HC - CH \\ \backslash \ / \\ N \\ H \end{array} \\
\begin{array}{c} H \quad H \\ HC - C - NH_2 \\ H \quad H \end{array} & \begin{array}{c} H \\ - C = NH \end{array} \\
\text{Ethylamine} & \text{Imines} \\
\\
\begin{array}{c} \qquad\qquad O \\ H \quad H \quad \| \\ HC - C - C - NH_2 \\ H \quad H \end{array} & \begin{array}{c} H \quad H \\ HC - CH \\ | \qquad | \\ O = C \quad C = O \\ \backslash \ / \\ N \\ H \end{array} \\
\text{Propionamide} & \text{Succinimide}
\end{array}
$$

Fig. 3.13 Nitrogen compounds. (See also Figs. 3.10, 3.21, and 8.4.) Not shown is the important constituent in living proteins, the amino acid, which has both an amine group and a carboxylic acid group on the same molecule (not ordinarily used in making ceramics). More discussion of the amines and amides in "DNA" will appear at the end of this chapter.

The simple amines have odors that are usually associated with long-dead fishes, although human lymph also has that odor. Since nitrogen usually has three bonds, it can make at least two directly to carbon atoms, and these are then called *imines*, instead of *amines*. There are several ways that imines can be configured, as shown in the diagrams. Some complex imines are used in ceramics as very effective dispersants.

If the OH were taken out of propionic acid and replaced by an amine group, an am*ide* would result, as shown in the figure. The proteins that are a large part of all animal flesh are composed of very complex amides. Wool, human hair, silk, and muscle are all amides. Some synthetic fabrics represent attempts to copy the natural amide chemical structure, which is held together partly by H-bonds.

An *imide,* as shown in the figure, combines the concepts of *imine* and am*ide.* Succinimide can be made from succinic acid, HOOC-C_2H_5-COOH, where the two OH groups at the ends are replaced by a single NH. Some complex imides are plastics whose high temperature capabilities rival those of ceramics for some applications, but with greater toughness and low cost. In a few electronics designs, imide films are used with ceramics to make highly optimized composites with a combination of useful properties.

3.2   Reactions

A Few Important Reactions. Just as hydrochloric acid will react with sodium hydroxide to make the chloride salt, acetic acid will react also, as shown in Fig. 3.14, to form the acetate salt, with water splitting off in each case. It should be noted that the organic groups are purposely written in different ways on the diagram, to accustom the reader to the many variations in the ways these might be encountered in the literature.

Instead of sodium hydroxide, ethyl alcohol will also react as shown, to form ethyl acetate, with water also splitting off. In this case, because the resulting product is all organic, it is not referred to as a salt, but instead it is called an "ester." Glycerin, with three hydroxyl groups per molecule, can react with three acid molecules to form an *ester,* which in this case is glyceryl trioleate. (The latter word is pronounced

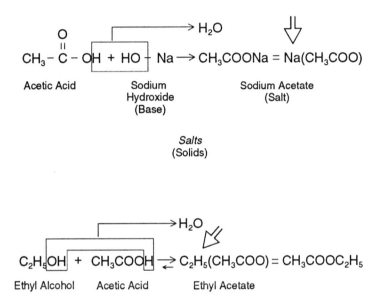

Fig. 3.14 The formation of salts and esters. A common name for glyceryl trioleate is "olein."

"try-ohl-ee-ate.") This is a naturally occurring liquid fat which is found in vegetables, fishes and to some degree in birds. As can be seen by the formula, it is "polyunsaturated," meaning that each molecule contains several double bonds (three in this case, that is, one on each oleate group). When eaten by humans, it is less likely to stimulate the body's production of cholesterol than is the saturated fat, glyceryl tristearate, which is found in the red meat obtained from animals. Because excessive amounts of cholesterol can tend to collect in the heart's arteries and eventually cause congestive heart failure, older people are becoming more aware of these relationships. Life insurance statisticians report that this sort of awareness and the resulting changes in eating habits have already increased the life expectancy of Americans. In addition to ester-stimulated cholesterol, the two glyceryl esters themselves also appear in human blood, and they are usually referred to by medical laboratories as "triglycerides." This is a somewhat illogical name, although it has come into common use, because it sounds as if there are three glycerins on the molecule, when in fact there is really only one, along with three acid groups. However, the attentive reader will note that this is not the only less-than-logical name in organic chemistry. Fortunately, the small mistakes in nomenclature made by early chemists are of little importance, while the benefits obtainable from their discoveries are of lifesaving magnitude.

The esterification reaction can be reversed by heating, usually with a catalyst such as a base. If glyceryl trioleate is heated with sodium hydroxide, glycerin and oleic acid will result, as shown by the small arrow in the figure. However, the sodium hydroxide will immediately react with the oleic acid, making the salt sodium oleate. This is soluble in water and is a very effective soap for cleaning dishes, hands, etc. It is the main ingredient in some liquid soaps, especially older formulations, and it is manufactured by heating vegetable oil with alkali. This process is called "saponification," which is related to the ancient Latin word "sapon," meaning soap. If glyceryl tristearate solid fat is saponified, a solid sodium stearate soap results, and that is the main ingredient of soap bars. In the nineteenth century, the widespread use of soap resulted in such a decrease in infectious disease that a great increase in population is said by historians to have resulted.

The saponification reaction done on a small scale in the analytical laboratory can be used to determine how much ester is present in a sample. Since the ester structure in glyceryl trioleate is necessary for its function as a dispersant in ceramic processing, a carefully controlled saponification reaction can be used as a quality control test for some types of fish oil dispersant, as will be brought up again in the chapter on dispersants.

Figure 3.15 illustrates a simple chemical reaction involving an organic compound, in which iodine is added directly to ethylene, with no water or other material splitting off. This reaction proceeds easily, without heat or catalyst being required. (For the definition of a catalyst, see Appendix II.) Other unsaturated organic compounds such as oleic acid also react easily with iodine, especially if they are both dissolved in alcohol to aid in mixing. In fact, this is a convenient test to determine whether a compound is unsaturated or not, and an example will be presented in the chapter on dispersants. The test can be used for quality control, wherever double bonds are needed for proper dispersing action, etc.

The next reaction in Fig. 3.15 involves two ethyl alcohol molecules (one purposely written differently from the other), in which a water molecule spits off and the organic remnants bond together during a reaction type that organic chemists call a "condensation." (This is a chemical process, and it does not involve the physical process of condensation that occurs when a liquid is boiled and then cooled.) The oxygen between two carbons gives the resulting compound its "ether" designation, just as in the MTBE structure shown in Fig. 3.8. This reaction is also called a "dimerization," in which two molecules of the same type of material come together and make a "dimer." (Mer is an ancient Greek word for thing, or unit.)

If a moderate number of such molecules bond together to form a medium sized molecule, the resulting product is called an "oligomer." (The political analog is that a single king rules by means of a monarchy type of government, but two or three dictators who rule together operate an oligarchy government.) If a very large number of molecules bond together, it is called a "polymer."

Polymerization. The bottom of Fig. 3.15 shows vinyl alcohol forming a *polymer*, via a reaction called "polymerization," where a

An Addition:

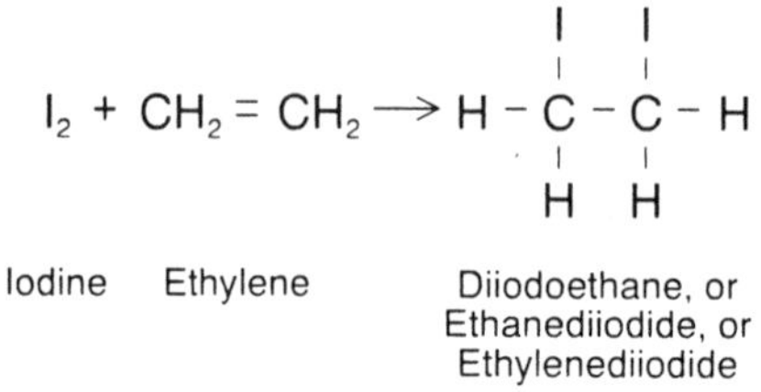

$$I_2 + CH_2 = CH_2 \longrightarrow H - \underset{\underset{H}{|}}{\overset{\overset{|}{|}}{C}} - \underset{\underset{H}{|}}{\overset{\overset{|}{|}}{C}} - H$$

Iodine   Ethylene   Diiodoethane, or
                                Ethanediiodide, or
                                Ethylenediiodide

A Condensation:

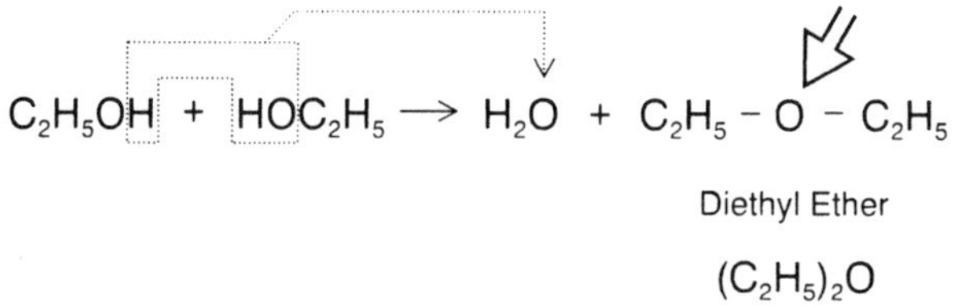

$$C_2H_5OH + HOC_2H_5 \longrightarrow H_2O + C_2H_5 - O - C_2H_5$$

                                                Diethyl Ether

$$(C_2H_5)_2O$$

A Polymerization:

Vinyl   Polyvinyl Alcohol
Alcohol   (a polymer)
(a monomer)   PVA

BINDER

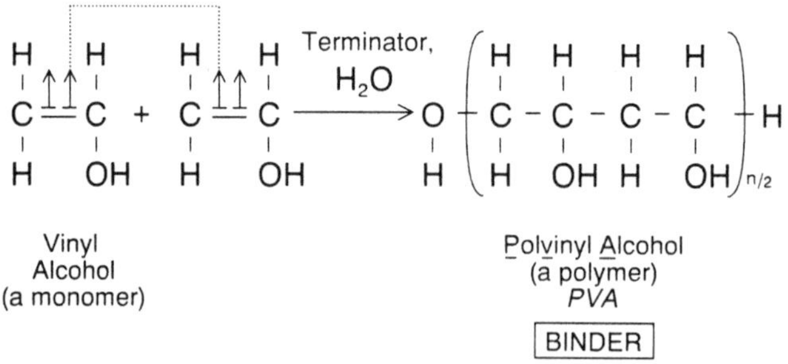

Fig. 3.15 Three organic chemical reactions.

large number of *monomer* molecules become bonded together. In this particular case, except for the ends of the chain, the reaction can go without any other material being added or splitting off; and only a rearrangement of the bonding electrons is involved. (Other types of polymerization can proceed by condensation in which something like water is split off, or by various other "mechanisms" involving hydrogen addition or removal, etc.) Each bond consists of a pair of electrons, and the topmost of the double bonds is drawn in the figure as if it is beginning to break apart into individual electrons. On the left-hand molecule, the right-hand electron swings around even further to the right and pairs up with an electron from the right-hand molecule to form a new bond and thus a dimer. However, on the same left-hand molecule, the left-hand electron is then available for bonding (it is temporarily a *free radical*), and it is very active in bonding to still another vinyl alcohol molecule, making a temporary trimer. In the same manner, the right-hand molecule bonds to whatever is nearby, making a tetramer, etc. The process can continue until all of the starting material is used up, with the resulting *polymer* involving literally billions of carbon atoms. The whole piece of material would then be a single, large molecule.

Because of the tangling effect, such a large molecule would most likely be a very brittle solid which is ordinarily not very useful. Therefore, organic chemists add a small amount of a "terminator" such as water in this case, which can stop the reaction at some desired lower level of average molecular weight. In the example shown by the diagram, a hydroxyl from the water will bond to one end of the polymer and a hydrogen to the other end. The resulting polymer is called *polyvinyl alcohol*, or *PVA*, which finds considerable use in ceramic processing as a binder. The more water present, the smaller the molecular weight that results, on the average. It must be realized, however, that this is a statistical process in which chance determines just when each particular chain will be terminated, and there is a great deal of random variation in chain length. If a certain small amount of terminator is fed into a stirred tank with a certain large amount of monomer, a desired *average molecular weight* will probably result. Ceramists should be aware that binders and other polymers tend to vary considerably from lot to lot during manufacture, and the quality of the ceramic product that uses these polymers can sometimes suffer as a result.

Since vinyl alcohol spontaneously polymerizes, it is not stable enough to store for a long time. Therefore vinyl acetate ester is made beforehand, and a small amount of water is later heated with it to convert it to vinyl alcohol.‡ Some of the residual water also serves as the terminator, leading to a desired average molecular weight.

3.3 Synthetic Polymers

Polyvinyls and Similar Thermoplastics. The vinyl-based polymers described in this section will all soften when heated. This category of polymers is referred to as "thermoplastic," meaning that they can be made to flow if hot. If the molecular weights are not too high, they will melt; however, if the "m.w." is too high and thus there is too much tangling to melt, then the polymers will become degraded with heat, sometimes returning to the monomer compound ("unzipping"), and sometimes becoming randomized "tars" or "chars," which are brown or black degradation products. Clean melting is desirable for injection molding, and clean unzipping is desirable for removal during the first stages of firing. Some polymers such as polyethylene do both, that is, they melt at one temperature without forming a tar, and they evaporate cleanly at a higher temperature without forming a char.

If the −OH of vinyl alcohol is substituted by hydrogen (situated on half of the carbons), that monomer would be ethylene (Fig. 3.1) and the polymer would be called polyethylene, even though there is no longer any double bonded ethylene present. Once again, the product is named after the starting material. Polyethylene is used as a binder in the injection molding of ceramics. Where only small numbers of monomer molecules have combined, and the n/2 shown at the lower right hand corner of Fig. 3.15 is only about 100, the *oligomer* is soft and melts easily, and it is usually called *wax*. The word *polyethylene* is usually reserved for the polymers where n/2 is around 1,000 or more, although there is no agreed upon definition. Polyethylene is one of the most heavily industrialized chemicals, and the quantities being used for

---

‡ The overall process usually starts with limestone, $CaCO_3$, which is heated to become lime, CaO. This is then reacted with carbon to beccome acetylene, $CH{\equiv}CH$, which contains a carbon-carbon triple bond. This readily undergoes an addition reaction with acetic acid, forming vinyl acetate.

containers and other applications are beyond nontechnical peoples' imaginations.

If the –OH of vinyl alcohol is substituted by chloride, the resulting polymerized compound will be polyvinyl chloride, usually abbreviated as "PVC" (not shown in the figures). This is only rarely used as a binder by ceramists, but it is very commonly used as artificial leather in automobile seat covers, as the insulating coating on electric wires, as packaging for food, and for many other purposes. It happens that vinyl chloride, the monomer, is one of the most carcinogenic materials known. Laboratory animals experimentally exposed to parts per trillion in the atmospheres of their cages almost all develop cancer. Vinyl chloride also evaporates readily. Therefore if any amount were known to be present in the polymers that literally surround us all, this would be a very dangerous situation. Luckily, the monomer is so extremely reactive that it all combines, and even the most sensitive measuring techniques have so far failed to detect any unreacted monomer in the PVC that we sit on and regularly eat from.

If the –OH of vinyl alcohol is substituted by an acetate group, –O–CO–CH$_3$, that product would be polyvinyl acetate (see also previous footnote). Figure 3.16 shows the compound that results when the OH is substituted by a carboxylic acid group, –CO–OH, and this is called polyacrylic acid. The dashes are added here to indicate the fact the group is positioned differently from the polyvinyl acetate situation. Of course, the oxygens of the carbonyls written here in the text as –CO– actually stick out at 90° angles, as shown in the diagrams. Also, most of the hydrogens are not shown in these particular diagrams.

If one of the hydrogens on the main carbon-carbon chain is substituted by a methyl group, poly*meth*acrylic acid ("PMA") results, and this can be used in ceramic processing as either a dispersant when the molecular weight is low, or a binder when the m.w. is higher. With a medium level m.w., that is, an *oligomer,* PMA can be used as a plasticizer. The methyl ester of PMA is PMMA, which is a common plastic (that is, a polymer) used to make hair brushes, jewelry, optical fibers, etc., as well as binders for ceramics.

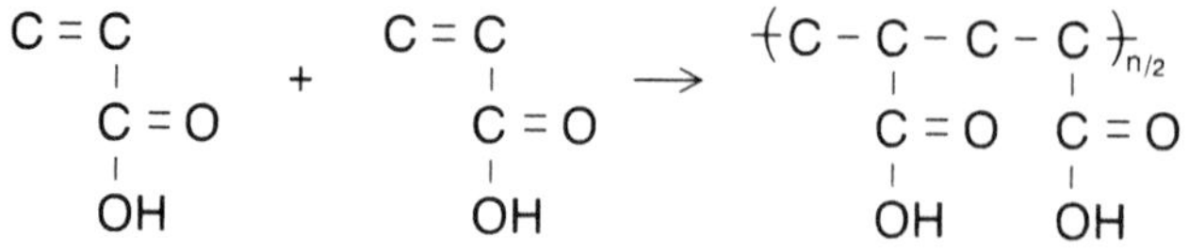

$$
\begin{array}{c}
C = C \\
\quad | \\
\quad C = O \\
\quad | \\
\quad OH
\end{array}
\quad + \quad
\begin{array}{c}
C = C \\
\quad | \\
\quad C = O \\
\quad | \\
\quad OH
\end{array}
\quad \longrightarrow \quad
\begin{array}{c}
{\large(}C - C - C - C{\large)}_{n/2} \\
\qquad | \qquad\quad | \\
\qquad C = O \quad C = O \\
\qquad | \qquad\quad | \\
\qquad OH \qquad OH
\end{array}
$$

Acrylic Acid   Polyacrylic Acid

                                                      ┌──────────────┐
                                                      │ DISPERSANT │
                                                      └──────────────┘

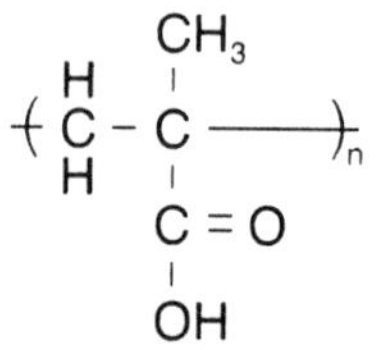

$$
\left(
\begin{array}{cc}
          & CH_3 \\
H   & | \\
C   & - C \\
H   & | \\
          & C = O \\
          & | \\
          & OH
\end{array}
\right)_n
$$

Polymethacrylic
Acid, PMA

┌──────────────┐
│ DISPERSANT, │
│   BINDER   │
└──────────────┘

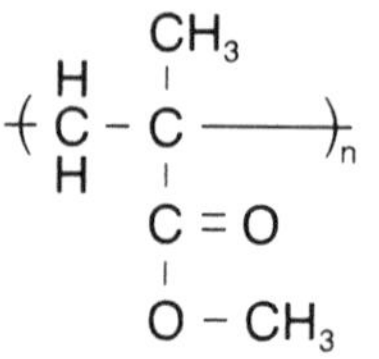

$$
\left(
\begin{array}{cc}
          & CH_3 \\
H   & | \\
C   & - C \\
H   & | \\
          & C = O \\
          & | \\
          & O - CH_3
\end{array}
\right)_n
$$

Poly(methylmethacrylate),
PMMA

┌──────────────┐
│   BINDER   │
└──────────────┘

Fig. 3.16 Polymerizations based on the double bond.

The simplest aldehyde (Fig. 3.4) is formaldehyde,[*] shown in Fig. 3.17. This can undergo a condensation reaction with PVA, resulting in one of the "formals," a reaction which is usually covered in college courses on organic chemistry. A reaction which is not so well known

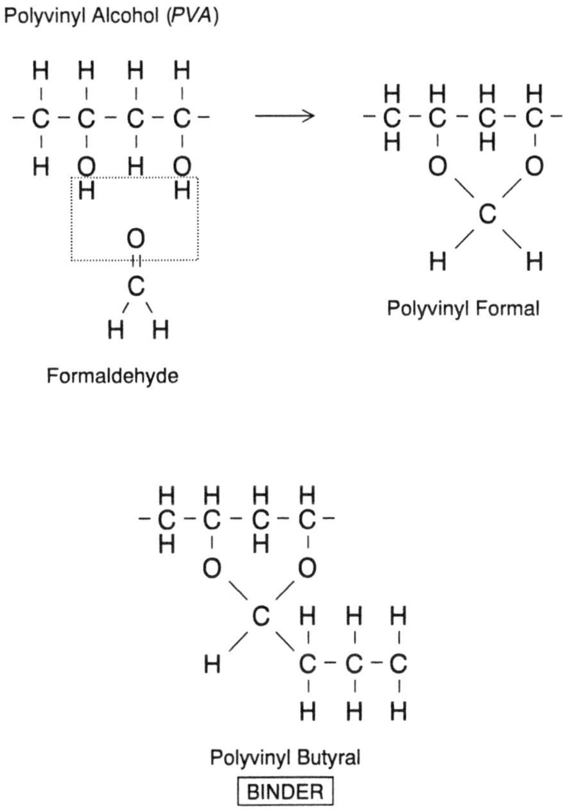

Fig. 3.17 Important PVA derivatives.

---

[*] Correspondingly, the simplest carboxylic acid is formic acid, HCOOH, which was not shown in Fig. 3.5. It is not commonly used in ceramics, although yttrium formate salt has been tried experimentally as a sintering aid.

but which is indirectly useful to ceramists is the formation of a "butyral," as illustrated at the bottom of the diagram. (Butyraldehyde was also illustrated in Fig. 3.4.) This polymer, which is often simply identified as "PVB," is a popular binder for making ceramics having many applications in the electronics industry. It is important to realize that, since PVB is made from vinyl acetate followed by polyvinyl alcohol in the supplier's factory, some residual acetate and alcohol groups are almost always present in the resulting commercial polymer used by ceramists. It turns out that these residues are quite important, as will be discussed in later sections, and some processes will not work without them.

Although vinyl alcohol spontaneously polymerizes, many other monomers such as ethylene do not, and they require either:

(1) another chemical (called a *curing agent*) to react strongly with them in order for the polymerization reaction to proceed, or

(2) *heat*, or

(3) a *catalyst.*‡

Either of the above three "initiators" of polymerization can then proceed further via either of the following two "mechanisms:"

(a) *free radical* reactions, as in the case of vinyls, or

(b) *ionic* reactions, as in the case of most condensations.

Ethylene glycol (Fig. 3.3) can be polymerized by a heat initiated ionic condensation to form polyethylene glycol (Fig. 3.18). Water is a terminator, which is used to make any desired approximate m.w. Again, there is no ethylene remaining in the product, or even any ordinary glycol, although the overall molecule is technically a very long glycol of sorts, since it has an OH at each end. It is sometimes referred to as a "polyether," because of the oxygen linkages. If the molecular weight is around 300, the material is a liquid, usually called "PEG-300," and it is used as a dispersant or a plasticizer in ceramics. However, if the m.w. is several thousand, this becomes a solid, and ceramists use it as a binder.

---

‡ For definition, see Appendix II.

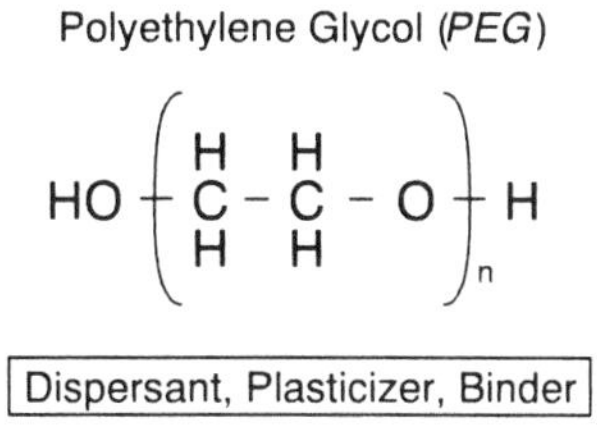

Polyethylene Glycol (*PEG*)

$$\mathrm{HO} \left(\begin{array}{cc} \underset{H}{\overset{H}{C}} & \underset{H}{\overset{H}{C}} \end{array} - O \right)_{n} H$$

Dispersant, Plasticizer, Binder

Fig. 3.18 The PEG family of compounds. Note that a mixture of n=6 and n=7 is a liquid that is commonly used as a plasticizer.

If the −OH of vinyl alcohol is substituted by a methyl group, the compound has the common name "propylene," and the systematic name "propene," as shown at the top of Fig. 3.19. When this polymerizes, it forms "polypropylene," sometimes called simply "polypro." This monomer polymerizes best in the presence of a catalyst, on the surface of which the individual molecules "adsorb" (defined in Appendix II) and then react with each other. The typical catalyst, such as zeolite mineral, has a surface which is quite rough on an atomic scale of sizes, although it might look smooth to the human eye. The microroughness has a repeating pattern which closely matches the shape of the monomer molecule that tends to stick to that catalyst surface (in other words, to *adsorb* onto the catalyst). As the adsorbed molecules polymerize and thus build up a "chain" of polymerized carbon atoms, other chains can sometimes form in parallel to the first one. These parallel chains also stick to each other, primarily by van der Waals bonding. The parallelism can be of good enough quality for the newly formed polymer to be crystallized, as shown in Fig. 2.3, in the drawing that is the third one from the bottom of the figure. The degree of crystallization is usually only very slight in most polymers, but it can be enhanced if the chain has something extending out from it in a regular manner, to mechanically force the carbons to line up. Polypropylene happens to be such a material, with the extra methyl groups extending out as they do. For this reason it crystallizes more readily than many other polymers, although only 20% or so of the polypropylene is actually crystalline in most cases.

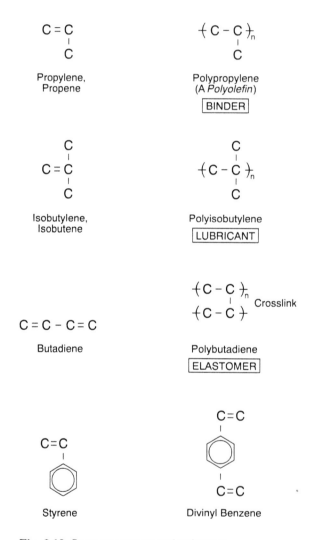

Fig. 3.19 Some monomers and polymers.

The "microcrystalline" regions can form in either of three ways: "isotactic," which is similar to *cis* in unsaturated molecules, or "syndiotactic," which is similar to *trans,* or "atactic," in which the methyls are randomly oriented. These have slightly different properties, but all are stronger and capable of higher temperature usage than are the usual non-crystalline polyethylenes. Polypro is used for making strong bottles, tubs, chemical equipment, and the like, where durability is important.

A relatively new development in polymers is that even polyethylene can now be made in highly crystalline form, reportedly more than 90% parallel. Special techniques are utilized, involving stretching to line up the chains, and various other proprietary inventions. This new plastic is still expensive, but it is extremely "tough." In other words, it is fairly hard because of its crystallinity, but unlike ceramics or even hardened metals, it can stretch a great deal before it breaks, so the total energy required to break it (that is, the toughness) is unusually high. It is used to make cables for racing sailboats, and it replaces steel cables but weighs much less.

Polyisobutylene can be seen in Fig. 3.19 to have two methyl groups extending outward, and these cause the melted material to have a high viscosity, because they interfere with flow. In the form of an oligomer that is liquid at room temperature, it is used to increase the viscosity of automobile motor oil, preventing the oil from leaking into the combustion chamber or out onto the ground. It has been used experimentally as a lubricant for dry pressing ceramic powders. When the m.w. is larger, the polymer is a strong solid, and this has been used experimentally as a binder which evaporates easily and therefore does not need oxygen to burn it out during the early stages of firing the ceramic.

A vinyl group attached to a benzene ring is called styrene, as shown at the bottom of Fig. 3.19. Polystyrene is quite brittle, because the large rings tend to restrain the sliding action that is necessary for untangling a highly stressed solid. An organic chemist might say the flow is "sterically hindered" in polystyrene, meaning that the three dimensional ("steric") shape restrains the flow. It is interesting that although polystyrene tends to crack too much for ordinary bulk uses, very thin sections can bend easily and do not crack under ordinary circumstances. This is analogous to glass fibers bending without breaking, while a thick glass rod could not be made to do that. Thin polystyrene finds everyday use as the styrofoam cellular plastic of discardable coffee cups. Because it is not very strong, polystyrene is not used much by ceramists.

Thermosetting Polymers. Straight chain butylene can have two double bonds, in the 1- and 3- positions, and this is called butadiene. (The lowest number would be chosen to identify the position of each bond, not 2- and 4-.) Organic compounds are practically never stable if two double bonds are on the same carbon, so there is no 1,2-butadiene, and therefore the numbers are rarely used. When butadiene polymerizes, most of the time it forms a straight chain, but a fraction of the time it sticks one end onto an existing chain and the other end onto another chain. When the monomer attaches two chains together, it is referred to as a "crosslink" (see Fig. 3.19). An extensively crosslinked polymer was previously shown diagramatically in Fig. 2.3. Other compounds having more than one double bond such as divinyl benzene (Fig. 3.19) can also act as crosslinking agents.

If the crosslinking is only present to a very small degree, the polymer chains sometimes twist into a helical, screw-shaped configuration, and this type of material can be stretched elastically over a fairly long distance without breaking. When released, it goes back to its original shape. Such a polymer is called an "elastomer," or a "rubber" (Fig. 2.3). Polybutadiene (Fig. 3.19) usually occurs in this form, and it is the most common type of synthetic rubber. Naturally occurring rubber is chemically somewhat similar, but it has to be crosslinked by heating with elemental sulfur. This fact was discovered by Charles Goodyear, and the sulfur crosslinking process is called "vulcanization," named after the ancient Roman god of fire and volcanoes, Vulcan. If the crosslinking is carried out to a greater degree, the polymer will become hard and brittle. In some cases this is desirable for the hardness aspect, but the toughness is generally greatest with a medium amount of crosslinking. (The natural protein helix is weakly crosslinked by H-bonds.)

Since the crosslinking process is usually initiated with the help of heat, further heating tends to make the polymer harder. This occurs instead of melting. Therefore these materials are called "thermosets," analogous to cement "setting up" or hardening, but carried forward in this case by a thermal effect. Some of these materials are capable of resisting moderate temperatures such as 300° C, and they therefore provide serious competition for ceramics, in many applications where low cost is important. Only a small amount of heat is needed to set them, compared to ceramics that require a great deal of energy for firing, so the cost of thermoset plastics is less.

The first large scale commercial plastic, which started a virtual revolution in manufacturing, was Bakelite®, shown in Fig. 3.20. This

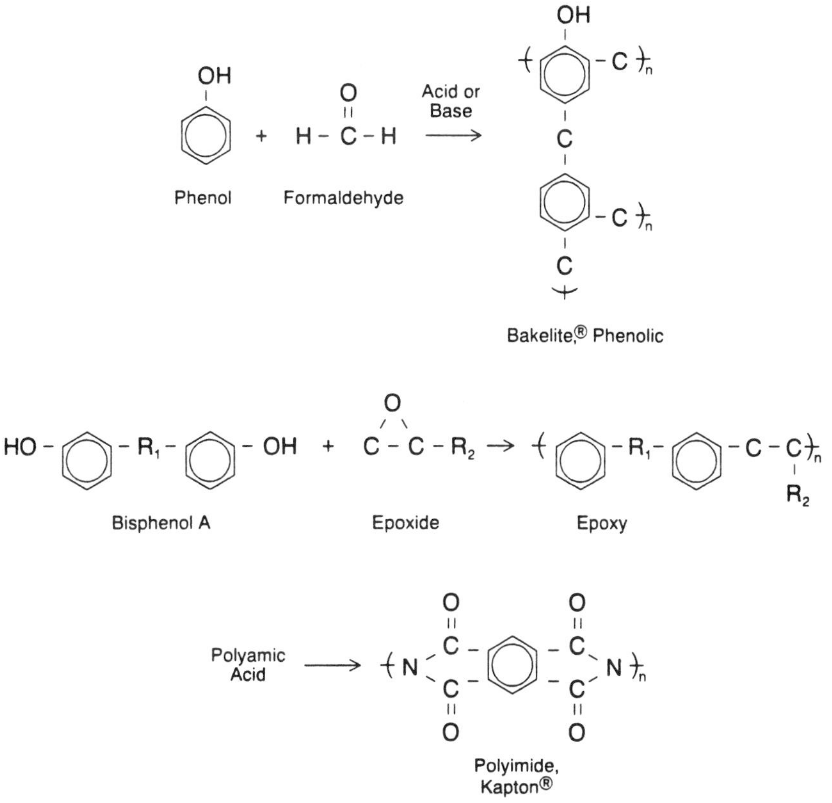

Fig. 3.20 Some thermosetting polymers. Polyamic acid is partly polymerized.

was invented in 1907 by Leo H. Baekeland. It is made by a condensation reaction between phenol and formaldehyde, and it is usually catalyzed by sodium hydroxide. It is highly crosslinked, and a good electrical insulator, but rather brittle. Because of its low cost, it replaced ceramics for many applications in the early days of electronic radios, telephones, etc. A major use is in gluing plywood lamination sheets together. It is often called "phenol-formaldehyde resin" or simply "phenolic." A *resin* is an old word for polymer, and it is not to be confused with "rosin," meaning the natural solid material which is extractable from pinewood sap. Phenolics are occasionally used as binders by ceramists in injection molding, when the ceramic is an oxide that can be fired in air for burnout. However, removal of the last traces of carbonaceous char is difficult, because further crosslinking continues until a high temperature is reached.

High Performance Polymers. Modern plastics have evolved from the original phenolics into many specialized offshoots, ranging from inexpensive materials such as the noncrystalline polyethylene used for packaging, to the highly crystalline but quite expensive polyethylene used for yacht cables. A moderately expensive and moderately high performance plastic which is now in very common use is the "epoxy," whose precursors are shown in the figure. Oxygen can go through an addition reaction with a carbon-carbon double bond, producing a slightly unstable and therefore active "epoxide" compound. This can act as a *curing agent* for a compound containing two phenol groups, and it can make a strong, highly adhesive plastic. When used as a thermosetting glue, the epoxy generates its own heat during "curing" (polymerizing). This makes the reaction proceed faster, which can cause vicious circle or runaway heating if the batch is too large, possibly even exploding. However, it is easy to control with small batching or forced cooling, and it finds widespread use, especially when reinforced by long glass fibers. In this extremely strong form, called "epoxy fiberglass," it is used to make chairs, roofing, boat hulls, etc. Because the epoxy chemically bonds to the high tensile strength glass fibers, excellent toughness is achieved.

Another thermosetting material, whose relatively high temperature capability makes it a rival to ceramics for some applications, is polyimide, shown in the figure. It is a good electrical insulator, and in thin sheet form it is used as a "substrate" or mechanical support for flexible printed circuits. Polyimide can be used up to about 300° C.

Among the expensive, high performance plastics, a fluorocarbon with the brand name Teflon® is particularly interesting because it was discovered by accident. The inventor was seeking a different fluorocarbon product and found a white polymer inside the apparatus which defied all attempts to clean it out. At that time, nothing would stick to it or wet it or dissolve it. Electrons are so tightly bound to the fluorine atoms that they are not easily polarized by electric fields, so van der Waals bonding to other materials is very weak, and this is the reason for its behavior. Nowadays it is used specifically because of those properties, for non-stick applications, low friction bearings, and resistance to chemicals. It was recently discovered that metal films can be made to stick to it by first treating the surface with liquid sodium metal to react with the fluorine. Teflon® keeps its strength up to about 250° C, although it does flow very slowly ("creep") at high temperature unless it is reinforced by chopped glass fibers. Other fluorinated polymers are also available, and ceramic slips are sometimes tape cast onto thin sheets of fluorocarbon plastics because of the easy release after drying.

If two carbon atoms are connected by a triple bond, the material C_2H_2 is called "acetylene." When hydrogen fluoride (HF) is added to it in a special manner, two fluorines bond to the same carbon atom, and "vinylidene fluoride" results. This can be polymerized to yield the plastic shown in Fig. 3.21, which is piezoelectric and thus competes with lead zirconate titanate ceramics for use in microphones, ultrasonic generators, etc.

Condensation reactions between alcohols and organic acids can lead to the formation of polyesters, which are used to make wrinkle resistant clothing, etc. If the alcohol is a short chain polyethylene with OH groups at each end, and terephthalic acid is reacted with this, the very strong plastic shown in Fig. 3.21 results. Thin sheets of Mylar® are used extensively in the ceramics industry as supports for tape casting slips. Electronic devices such as capacitors, multilayer ceramic packages, etc. are mass produced in this manner.

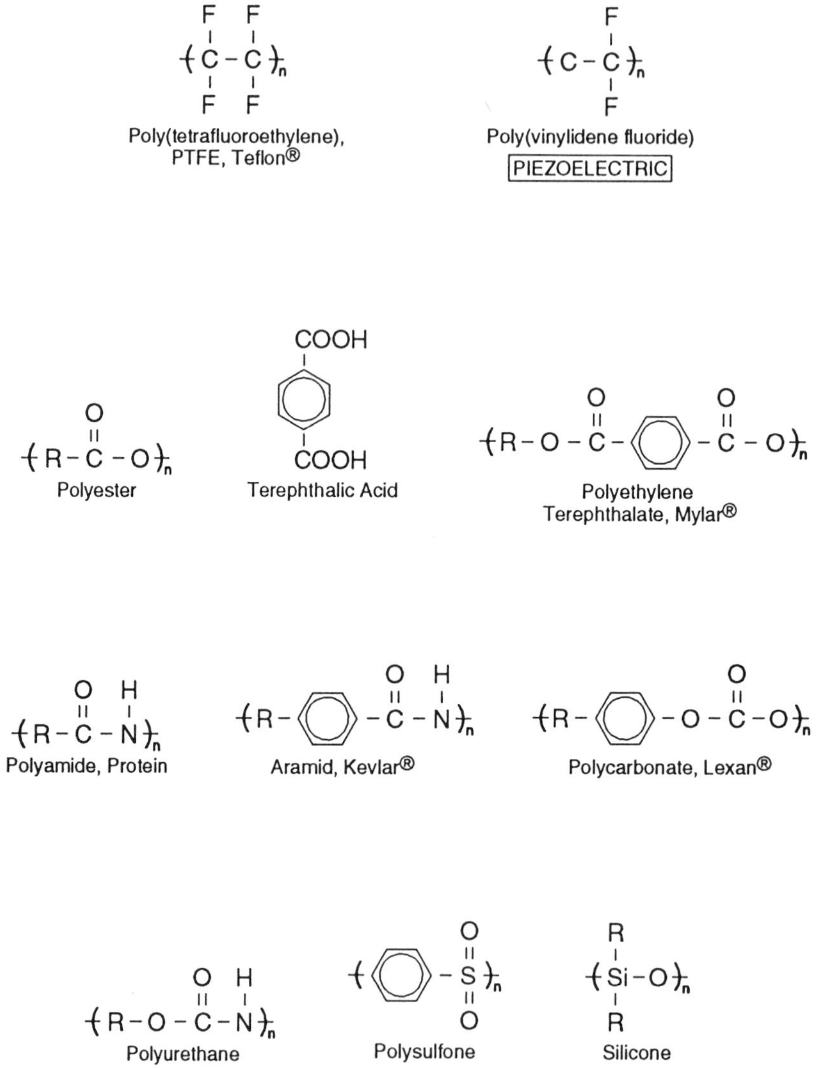

$$\left(\text{C}-\text{C}\right)_n$$

Poly(tetrafluoroethylene),
PTFE, Teflon®

$$\left(\text{C}-\text{C}\right)_n$$

Poly(vinylidene fluoride)

| PIEZOELECTRIC |

$$\left(\text{R}-\overset{\text{O}}{\underset{}{\text{C}}}-\text{O}\right)_n$$

Polyester

COOH

COOH

Terephthalic Acid

$$\left(\text{R}-\text{O}-\overset{\text{O}}{\underset{}{\text{C}}}-\bigcirc-\overset{\text{O}}{\underset{}{\text{C}}}-\text{O}\right)_n$$

Polyethylene
Terephthalate, Mylar®

$$\left(\text{R}-\overset{\text{O}}{\underset{}{\text{C}}}-\overset{\text{H}}{\underset{}{\text{N}}}\right)_n$$

Polyamide, Protein

$$\left(\text{R}-\bigcirc-\overset{\text{O}}{\underset{}{\text{C}}}-\overset{\text{H}}{\underset{}{\text{N}}}\right)_n$$

Aramid, Kevlar®

$$\left(\text{R}-\bigcirc-\text{O}-\overset{\text{O}}{\underset{}{\text{C}}}-\text{O}\right)_n$$

Polycarbonate, Lexan®

$$\left(\text{R}-\text{O}-\overset{\text{O}}{\underset{}{\text{C}}}-\overset{\text{H}}{\underset{}{\text{N}}}\right)_n$$

Polyurethane

$$\left(\bigcirc-\overset{\text{O}}{\underset{\text{O}}{\text{S}}}\right)_n$$

Polysulfone

$$\left(\overset{\text{R}}{\underset{\text{R}}{\text{Si}}}-\text{O}\right)_n$$

Silicone

Fig. 3.21 Common high performance polymers.

It has already been mentioned that protein, hair, muscle, and many other materials in animals are polyamides. Synthetic variations on this structure include plastics such as Nylon®, Orlon®, etc., which are imitations of silk and wool. If a benzene ring is included in the structure, the extremely tough aramid resin results, which is named after the aryl benzene ring plus the amide linkage. This is used to make armor, racing car bodies, etc., where the effective strength must be at a maximum (sometimes used as composites with ceramics). However, like many of the high performance polymers, the cost is presently quite high.

Polycarbonate is another tough material, and for this reason it is used to make shatterproof windows as a replacement for glass in school buildings, etc. It is also highly transparent to light, so it is used for compact disk substrates and optical fibers, in which is competes with silica glass. A new grade of polycarbonate unzips to the monomeric form when it is heated, and it is being used as a binder for ceramic compositions where firing can not be done in air, as with aluminum nitride. It is also a binder in powder metallurgy, particularly in the manufacture of sintered tantalum capacitors.

Polyurethane is a relatively simple polyimide, whose toughness leads to use in floor coatings. For ceramic processing it is used as a non-wearing coating on the inside of steel ball mills. Because it is resilient, it resists being chipped off by the grinding media or powder. However, it can be swelled up by organic solvents such as toluene, and then it is weakened and can become incorporated into the slip, resulting in large pores when it is eventually burned out. It is best used for this purpose with water or isopropanol based slips, which are too polar to swell the non-polar polyurethane.

Although most thermoplastics soften at temperatures below about 100°C, polysulfone is unusual in that it is a true thermoplastic and therefore can be injection molded like polyethylene, but it softens at about 300°C. Silicone* is also useful up to about 300°C. Some low m.w. silicones are liquids, and some are slightly crosslinked** rubber ("room temperature vulcanizing," or "RTV," since they polymerize readily). Some are brittle solids. A special silicone formulation has organic groups that split off easily when heated, and a 400°C heat

---

* The (-Si[R_2]-O-) is called a "siloxane group." If one of the R radicals is an -OH, then this group is called a "silanol," often found on the surface of silica, etc.
** Crosslinking is sometimes called "curing" (or "vulcanizing" in rubber).

treatment converts this to a silica-like glass. This is referred to as "spin on glass" in the semiconductor industry, because a thin layer of it is applied evenly to silicon wafers by putting a drop in the middle of the wafer, and then spinning this to sling off most of the silicone by centrifugal force. After heating, it contains a small amount of organic material, so it is not pure silica. However, it makes a good electrical insulating layer, and it is tougher than silica.

Copolymers. Two different monomers such as PVA (Fig. 3.15) and PMA (Fig. 3.16) can be polymerized together, and the resulting material is called a "copolymer." It can then have two or more "functional groups" on the molecule, such as OH and COOH, each of which can perform different functions. One part of the molecule might chemically combine with a certain type of site on a powder particle (possibly acidic), and the other part of the molecule might combine at a different site (possibly basic). A copolymer of this kind, having both basic and acidic (or other) groups at various places on the molecule is called "amphipathic." This word is analogous to the word "amphoteric," applied to ionic compounds such as aluminum hydroxide that can be both basic and acidic. *Amphipathic* copolymers are sometimes useful as dispersants and binders in ceramic processing, because they chemically combine with powders at more than one type of site on the powder surface.

The two different kinds of monomer (such as A and B) can be polymerized either as a "random copolymer" with a configuration AABABBAAABABA, etc, or a "block copolymer," with AAAAABBBBAAAAABBB, etc. The block copolymer type is most commonly used when amphipathic properties are desired.

Characterization. One of the most important characteristics of any polymer is the molecular weight, or m.w. Looking at the structure of PEG in Fig. 3.18, for example, each $-C_2H_4O-$ unit inside the parentheses is called a *mer*[†] unit, with a molecular weight of 44, not including the H or OH terminating groups. If n, the number of mers, is 6, then the m.w. of the whole oligomer molecule, including the terminating groups, is 282.

[†] "Mer" was an ancient Greek word meaning, not "molecule" of course, but "thing."

When a chemist is simply handed a beaker of oligomer or polymer without being told the value of n and asked to determine the molecular weight, there are several experimental methods for doing this. For example, a solution of the polymer can be diffused through an inert gel, in which case the smaller molecules will diffuse fastest. This is called gel permeation chromatography, or GPC. In almost all examples with commercial polymers, there will be more than one molecular weight, and the resulting "m.w. distribution" will include a fairly wide range of numbers. A typical example might be something like "8% below 1000, 13% between 1000 and 2000, 41% between 2000 and 3000, 29% between 3000 and 4000, and 9% between 4000 and 5000." The GPC yields a "number average molecular weight," which is the conventional way to calculate an overall figure when a wide distribution is involved. In other words, with the above example, the *number average molecular weight* would be

Relative Amount	Median M. W.	
0.08	x 500 =	40
0.13	x 1500 =	195
0.41	x 2500 =	1025
0.29	x 3500 =	1015
0.09	x 4500 =	405
Totals: 1.00		2680 / 1.00 = 2680 # avg. m.w.

Another practical method for measuring the m.w. is light scattering, in which the larger molecules scatter more ultraviolet light than the smaller molecules do. However, the large ones actually scatter much more than the small ones. This method yields a "weight average molecular weight," where the larger molecules count much more than the smaller ones; in other words, it is a "weighted average." If the m.w. distribution in a certain sample happens to be somewhat rich in the larger molecule sizes, the *weight average molecular weight* determined by light scattering will be a much larger number than if GPC had been used to measure a number average molecular weight.

Several other methods for measuring the m.w. are occasionally used, including viscosity, osmosis, and sedimentation with an ultracentrifuge.[7] Whenever the m.w. is specified by a supplier of binders, etc., care should be taken to determine whether it is the number average or the weight average that is being quoted, in order to properly compare the material to similar products from other suppliers.

Another important characteristic of a polymer is the "glass transition temperature," or T_g. This is measured by a method that polymer scientists call "thermomechanical analysis," or TMA. (Ceramists would most likely call this "dilatometry.") The height of a polymer sample is measured continuously with a mechanical probe resting on its top surface, while it is heated slowly. The thermal expansion is thus determined. Typical results are shown in Fig. 3.22. Below the T_g, polymer scientists refer to the material as being "glassy," because it is relatively hard and brittle. Although most polymers are somewhat amorphous (essentially non-crystalline), a certain degree of crystallinity can be present in this temperature region, and the word is left

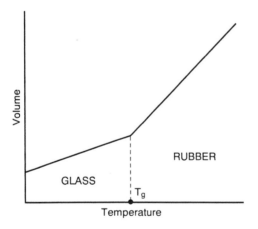

Fig. 3.22 Thermal mechanical analysis (TMA) and the glass transition temperature.

---

[7] A. W. Adamson, "Physical Chemistry," Academic Press, New York (1986) 372, 920.

over from an earlier period in scientific history when the microcrystallinity of polymers was not measurable. Above the T_g, polymer scientists refer to the material as being "rubbery," because it is relatively soft and elastic. However, most polymers are not capable of the great elongations that true helical rubber structures such as polybutadiene can easily undergo without breaking. Therefore these are relative and mainly historical terms. (It should be noted that inorganic silica based glasses have similar expansion characteristics and a T_g, although there is much more brittleness in both regions.) From the standpoint of the ceramist who uses a polymer as a temporary binder for an inorganic powder, the importance of the T_g is that the polymer-powder *green* composite is somewhat more brittle below that temperature than above it. At room temperature, a liquid which polymer chemists call a *plasticizer* (mentioned briefly in earlier sections) can be added to the binder in order to lower the T_g thus moving the whole material into the less brittle "rubber" region of the diagram. Being less brittle, the green material is less likely to break accidentally during later processing steps (inspection, storage, firing, etc.).** More use of this concept will be made in later chapters.

Materials scientists do not consider the *glass transition temperature* to be a true "first order phase transition." That is, the "glass" and "rubber" are not different "phases" in a thermodynamic sense. If the heat content, or enthalpy, is plotted versus temperature, the graph looks like the one in Fig. 3.22. Below the T_g, the slope is the heat capacity at constant pressure, C_p, which is the first derivative of H versus T. In this lower region of the plot, the first derivative is essentially constant. Above the T_g the slope has a different value, which is also essentially a constant. Right at the T_g, the slope changes, but there is no sudden increase in heat capacity (in order to break physical or chemical bonds), as there would be in a true *phase transition* such as physical melting or some kind of chemical transition. That is, the C_p does not suddenly jump to a high value (the latent heat of fusion, or etc.). However, it

---

** A polymer with a T_g below room temperature is equivalent to being self-plasticizing, and it often will not need the addition of another material to make it less brittle. An example of such a material is Rohm and Haas poly(oxazolidinyl)-ethyl methacrylate.

would have gone through a sharply discontinuous increase if the phase had been truly changing at that point.

Materials scientists do, however, refer to the T_g as a "second order phase transition." (This term is frequently mentioned in polymer science literature but is not usually explained.) The second derivative of H versus T, which is the change in slope, is essentially zero going up the line toward the T_g, since the slope is quite constant. However, the second derivative is instantaneously high right at the T_g because the slope changes there. The second derivative then goes back again to essentially zero while going along the constant-slope line above the T_g. The large instantaneous value of the second derivative is the reason why the term "second order transition" is used. It has a jump to a high value, just as the first derivative would have during melting, etc. (The experimental data obtained from a TMA done on a commercial polymer sample is usually not as straight-line linear as the simplified diagram in Fig. 3.22.)

3.4 Natural Carbohydrates

Figure 3.23 shows a molecule of "glucose," one of the "carbohydrates" consisting of carbon, hydrogen, and oxygen, which occur very commonly in both plants and animals. Polymers of *glucose*, along with the amide polymers of protein, are the main building blocks of life. Glucose can form the ring structure shown in the diagram, as follows: one of the electrons of the OH's electron-pair swings over to pair up anew with one of the electrons of the aldehyde's oxygen double bond. This makes a six-membered ring, which tends to be stable because the carbon-carbon bond angle (109°) is nearly that of a geometric hexagon (120°), as discussed in earlier sections (Figs. 2.4 and 3.6). The natural bond angle of the oxygen is approximately 90°, so the ring is not really flat (that is, it could not be in the plane of the paper, as shown in the figure), but instead it is twisted somewhat to accommodate the various bond angles. Two glucose monomers can undergo a *condensation* reaction, losing water, to become a *dimer*. This reaction can then continue further, polymerizing literally billions of glucose units into a long chain molecule.

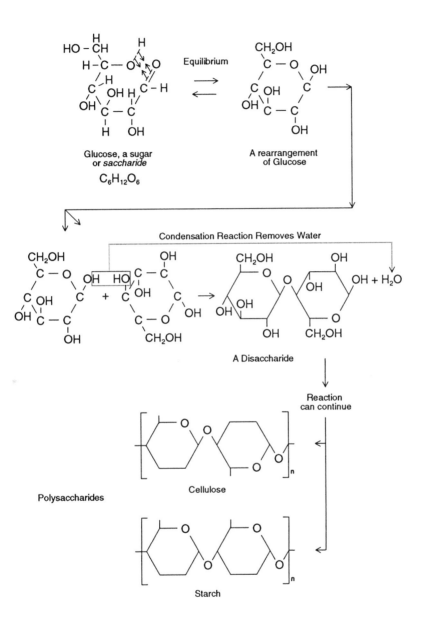

Fig. 3.23 Glucose polymerization reactions.

The dimer can form in either of two ways, each of which is equally stable. Without a solid model to hold in your hand and rotate at various angles, those two different materials, "cellulose" and "starch" do not appear to be very different. However, there is a functional difference which is very important to life. In *cellulose,* the OH groups of nearby chains can easily form hydrogen bonds to each other. Since it is strongly bonded to itself, cellulose is therefore insoluble in water and also mechanically strong. Because of these two properties, it is the material used by nature to give wood and other plant materials their structural strengths. Tubes of cellulose are the fibers that carry water up from plant roots to the leaves, via capillary action. Of course, if they were soluble in water, they would not be usable for this purpose. From the standpoint of strength, cellulose fibers are analogous to the fiber-glass reinforcements in man-made epoxy structures such as chairs and tables used in schools, offices, etc. Although cellulose is not directly soluble in pure water, it can be transformed into soluble byproducts by various strong chemical agents. A few animals are able to do this in their digestive systems in order to use cellulose as food, but humans and many other animals can not. Although we humans have not been able to make practical foods out of cellulose, we have succeeded in dissolving it by special means for other uses. For example, it can be dissolved in strong nitric acid or "copper-ammonia complex" basic solutions and later precipitated in the form of long fibers, marketed as Rayon® imitation silk cloth.

Starch is oriented just differently enough so the OH groups can *not* easily form hydrogen bonds. Therefore, the chains can be pulled away from each other, and starch does easily dissolve in water, unlike cellulose. Once in solution, the water forms hydrogen bonds all around the chain, since the water molecules are free to move into whatever is the ideal configuration for bonding.* Starch is used by plants to store energy and necessary chemicals, often for later use by germinating seeds.

---

* It should be noted that the water molecules were previously H-bonded to each other, so those previous bonds had to be broken in order for the new ones to form between the starch and the water. This would not happen unless the new bonds to the starch were more energetic than the old ones between the water molecules. This point will be brought up later in the chapters on solvents and binders.

It is also used by mankind in industrial applications as a cheap but weak glue, and by ceramists as a cheap but weak binder for bricks, etc.

Chemists have transformed natural cellulose into a soluble, "semisynthetic" compound by taking off the hydrogen from one of the hydroxyl groups and replacing it with a methyl group, as shown in Fig. 3.24. This material, methyl cellulose (sometimes written as a single word), has a delicate balance between H-bonding to itself and competitively H-bonding to water. Therefore a small change in temperature can cause it to either dissolve or be solid. In fact it, after it dissolves in water at room temperature, it can be made to gel (become a solid network while still intermingled with the water molecules) if it is then slightly heated.[8] This unusual property, called "thermal gelation," is used by ceramists to stiffen freshly extruded green bodies by the application of heat, in order to prevent them from collapsing as they come out of a complex extrusion die. The water can be evaporated in a stream of hot air, and after it is gone, the green body will become solid simply because it is then completely dry.

In the middle-right hand diagram of Fig. 3.23 it can be seen that, after the end hydroxyl groups have linked up to neighboring rings, becoming *ether* oxygen linkages instead of pairs of hydroxyls, there still are three hydroxyls remaining unreacted on each ring. The OH on the methyl group that sticks out is the first one to be substituted in methyl cellulose, becoming a methoxy group (Fig. 3.24, upper-middle diagram), but further "methylation" can still take place (not shown in any of the figures), until all three oxygens have become methoxy groups (but not any of the ether oxygens). Whether one, two, or three methyls have been added is described by the term "degree of substitution," or "DS," which of course is either 1, 2, or 3. In specifying which grade of methyl cellulose (or the other substituted materials in Fig. 3.24) is being sold by a supplier, usually the DS and the m.w. are both listed.

If a carboxyl group, COOH, is put on the methyl, the resulting material is a weak acid, carboxymethyl cellulose or CMC (see figure again). A sodium salt can be made from it, and this material is even more soluble in water than methyl cellulose. When large amounts of it

---

[8] N. Sarker, et al., Am. Ceram. Soc. Bul., 62 (1983) 1284.

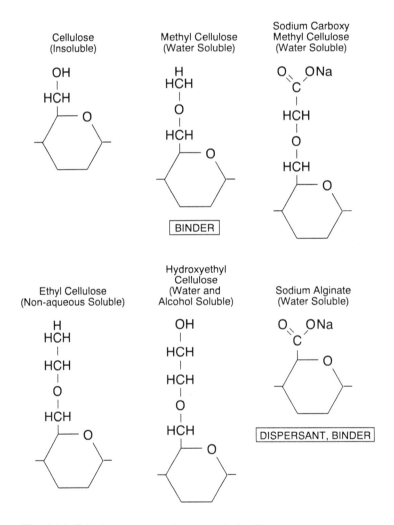

Fig. 3.24 Cellulose, compared to some derivatives.

are dissolved, the viscosity is fairly high, with a "plasticity" effect (in a ceramics sense) whereby a medium amount of force will make the solution flow easily. For this reason, sodium CMC is used in food, water based paint, etc., to control flow properties. Pure grades are safe to eat, although they have no actual "food value" because they are not metabolized by the human body. However, they can cause food products like mustard, catsup, and salad dressing to flow when poured vigorously, but they will not "run" off the edge of the main food under only the influence of gravity. Similarly with paint, the solution will flow when brushed but will then not drip down the side of a vertical wall under gravity alone. When this material is used for ceramic processing, it leaves sodium oxide behind when it is burned out, and for this reason it is not used a great deal in ceramics, except in making tile. However, there is not much sodium in the total composition, compared to the impurities present in many ceramics, and also a large percentage of the sodium oxide actually evaporates at typical ceramic firing temperatures, so CMC should possibly be considered as a combined dispersant/binder more often.

If an ethyl group is put on the cellulose molecule instead of the methyl, becoming ethyl cellulose, this structure is just nonpolar enough so it will dissolve in alcohol and other moderately nonpolar organic solvents (such as low m.w. PEG, etc.). This is used as a binder for screen printing inks, especially in producing ceramic-metal composites ("cermets") for electronics.[9] When a large amount is dissolved, it can provide plasticity just like the other *semisynthetic* cellulose compounds do. In general, the materials made by slightly modifying cellulose (in other words, the "cellulosics") can almost all be used to raise the viscosity of solutions, either aqueous or nonaqueous, often providing some degree of plasticity for *non-plastic* powders such as coarse alumina.

To further modify the molecule, an OH group can be put on the end of the ethyl group, and the resulting material is soluble in both alcohol and water. (These various solubility relationships are not entirely obvious, and they were originally determined more by experimentation

---

[9] J. B. D'Andrea, U.S. Patent 2,924,540 (1960); L. F. Miller, "Thick Film Technology," Gordon and Breach, New York (1972); D. J. Shanefield, page 284 in "Ceramic Films and Coatings," Noyes Public'ns., Park Ridge, NJ, J. B. Wachtman, Jr., and R. A. Haber, eds. (1993) [see p. 292].

than by theoretical reasoning.) This compound is sometimes used for a variety of applications instead of the other *cellulosics*.

Some naturally occurring water soluble materials are obtainable from several plant products, usually by minimal alkali or fermentation treatment, followed by filtration to separate the insoluble parts of the plant. A type of seaweed called kelp can be treated with dilute sodium hydroxide to yield the "alginate" compound shown in the figure. (The structure is actually more complex than the simplified version shown in the figure.) Strictly speaking, kelp is not a type of algae, but like many such older chemicals, it is slightly misnamed. Alginates are used extensively in food and also as glue for envelopes, stamps, etc. For ceramics, the sodium salt is both a dispersant and a binder. It is inexpensive and fairly powerful for both purposes, but it does not burn out easily, and it tends to crosslink and leave a thermoset tar unless a great deal of air is available for burnout. However, this is true of the other cellulosics, as well.

Another natural product which is both a dispersant and a binder is guar gum, also shown in Fig. 3.25. In this case the hydrogen of the OH group in cellulose has been replaced by a whole six-membered ring. That interferes with H-bonding to itself, and so the structure (actually more complex than shown) is water soluble. This compound has an advantage in that it can be chemically gelled instead of thermally gelled, by spraying a salt of boric acid onto the extruded or cast green body immediately after shaping. Since that would leave a residue of boron oxide after burnout, which might change the properties of the fired ceramic, completely burnable organic chemicals for gelling the gum could be sprayed on instead, including epichlorohydrin or the dye[10] "congo red." Again, guar gum is used to some degree as both a dispersant and a binder for ceramics. Some varieties burn out easily. Guar gum is made by refining locust beans that have been fermented. This is an ancient art, originally developed in northern India where locust beans are a native plant, and its earliest uses were probably for making paints and cosmetics.

---

10 G. Y. Onoda, Jr., in "Ceramic Processing Before Firing," G. Y. Onoda, et al., Eds., J. Wiley, New York (1978), 235 [see particularly page 248].

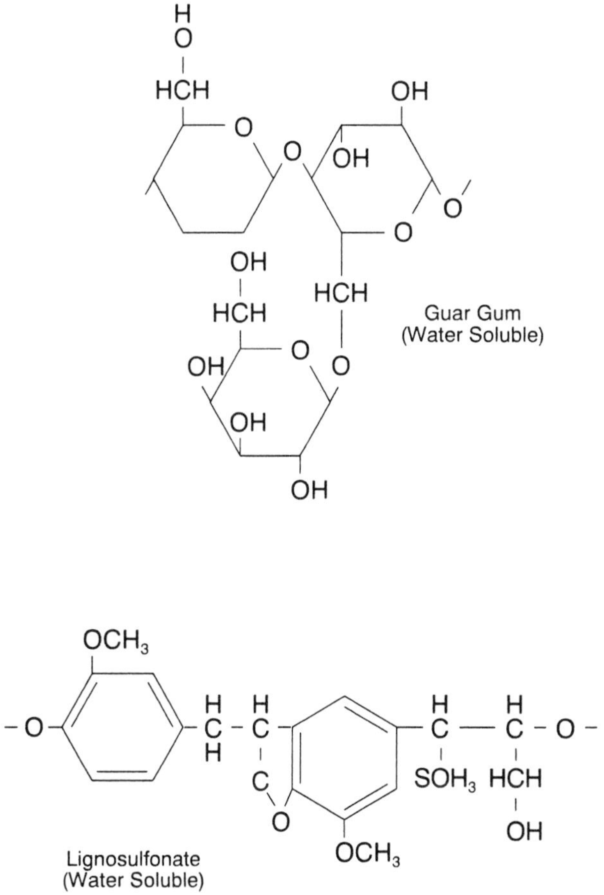

Fig. 3.25 Simplified diagrams of typical structures occurring in a gum and a lignin derivative. See discussion of Fig. 8.5 for more background on sulfonates.

Many other natural gums are used occasionally by ceramists, often borrowing very old ideas from the makers of paint, ink, cosmetics, and glue. An example is gum arabic, which is sometimes called acacia or gum acacia. At one time the author was the responsible engineer on a project involving the large scale production of ferrite magnetic ceramics. The dispersant used was gum arabic, which had been used successfully for many years previous. A yearlong lack of rain in the Middle East caused the available gum arabic to become chemically altered to the point where it was no longer usable, and no other dispersant seemed to work properly in this particular ceramic process. Therefore the author had to work nights and weekends to help develop a different process. Unfortunately, this sort of thing sometimes happens with natural products, and that is one reason why modern ceramists tend to use synthetic materials whenever possible. However, some natural products have such useful properties at low cost that it is difficult to find synthetic substitutes, and many of these products are therefore still used extensively.

Since the six-membered rings in the cellulosics do not have alternating single and double bonds between their carbon atoms, these rings are not *aromatic* or true benzene-type groups. However, benzene rings do occur in plants and animals (as well as in petroleum and coal). For example, they are present in certain amino acids which make up proteins. Also, they are present in lignin, the material that acts like a matrix glue to hold the vertical cellulose tubes together in the wood of a tree trunk. Lignin is analogous to the epoxy in man-made fiberglass reinforced plastic chairs, tables, etc. In order to make paper out of wood, ground up sawdust is partially dissolved in sulfurous acid (H_2SO_3), and the lignin is chemically modified to become "lignosulfonate," as shown also in Fig. 3.25. This is water soluble, and its solution is mostly discarded, while the insoluble cellulose is then filtered off to be processed into paper. A portion of the lignosulfonate, whatever the market can absorb, is used to make other products such as a dispersant and/or binder for the ceramics industry. It is not a strong binder and it does not burn out well, but being an otherwise discarded byproduct, it is inexpensive. When it is chemically refined into certain lower molecular weight ranges, it can be an excellent water based dispersant, and it is now used in increasing quantities to manufacture tile and other ceramics. For this purpose it is ordinarily sold in the form of the sodium or ammonium salts, to make it easier to dissolve the materials in water.†

The lignosulfonate molecule is actually more complex than is shown in the figure. It consists of many different variations, one of which has an additional oxygen between the sulfur and the carbon atoms ("lignosulfate"). Some interesting dispersion properties have been reported, as will be discussed later.

3.5 Hydrogen Bonds and Life

At this point in the discussion, the reader should be able to appreciate that life as we know it on earth would be impossible without the existence of the H-bond, because it is a fairly weak type of bond, but not too weak. This is explained further as follows. Living things depend on the principle of "replication," the process by which many new protein molecules can be formed in the same shape of a single old protein molecule. Several different "amino acids" (see caption of Fig. 3.13[**]), which are the building blocks that make up any new protein molecule, become bonded simultaneously to the surface of an old protein molecule. These small amino acid molecules are held onto the old surface by hydrogen bonding. The small amino acids then bond to each other, end to end, by *amide* linkages (Fig. 3.13), making a new, long protein molecule, so the protein is strongly (covalently) bonded within itself, but only weakly (H-bond) attached to the old, long protein. This attachment must be weak enough so that the newly formed protein molecule can then easily break away and make room for still another new one to form, without damage. H-bonds are weak enough for this and yet are strong enough for the process to go forward again the next time.

---

† It is a general principle that acids that are fairly weak, because they do not ionize very much, do not dissolve easily in aqueous systems. However the sodium or ammonium salts are fully ionized, even in the dry solid state, and so they quickly dissolve in polar solvents such as water.

** The simplest *amino acid* is "glycine," which is $(NH_2)CH_2(COOH)$. It is both a base and an acid, and it can easily H-bond to many other types of molecules.

The long protein molecule is spiral-shaped, and when two of these spirals are weakly H-bonded together, side to side, this is the well known "double helix" structure.   In general terms, that is the configuration of the genetic code protein, deoxyribonucleic acid,[1] abbreviated "DNA."

4. CERAMIC PROCESSING FUNDAMENTALS

4.1 The Process Steps from Powder to Ceramic

The author studied chemistry and chemical engineering in school but never actually studied ceramics as an academic discipline. Therefore this book is written from a somewhat unusual perspective, more like a chemical engineering view of processing than a typical ceramics view. The emphasis on a certain few aspects of processing is based on those serious problem areas that became apparent during the author's several decades of industrial experience, rather than being based on theoretical considerations.

At the outset of this discussion, it should be noted that several adequate books[1] are available for further study of ceramic processing in general. For that reason, only those few key aspects of processing that interact strongly with organic additives will be explained in detail here.[†] Many of these process steps are also used in powder metallurgy, ink, and paint technologies.

---

[1] (a) W. D. Kingery, et al., "Introduction to Ceramics, J. Wiley, New York (1976); (b) J. S. Reed, "Principles of Ceramics Processing," J. Wiley, New York (1995) ; (c) D. W. Richerson, "Modern Ceramic Engineering," Marcel Dekker, Inc., New York (1992); (d) R. H. Perry and W. D. Green, "Perry's Chem. Eng'r's. Handbook," McGraw Hill, New York (1992) [Basic processes such as drying, filtering, pressing, etc.]; (e) J. T. Jones and M. F. Barard, "Ceramics: Industrial Processing and Testing," Iowa State Univ. Press, Ames (1993); (f) Chapters on ceramics appear in the materials science books listed in Appx. III.

[†] Chapter ten lists specific additive formulations, processing conditions, and typical problems with their recommended solutions. The interested reader should see particularly page 288 for such case study problem solving, and one should be aware that many processes other than the "tape casting" method described in the table will share the same types of problems (footnote continued on next page)

The Powder. A chemical engineering view of a typical ceramics manufacturing process is shown in Fig. 4.1, in simplified "schematic" form. The process usually starts with a powder, examples being naturally powdered clay, or artificially powdered alumina, barium titanate, etc. A powder has a relatively large *specific surface area,* or surface area per given weight of material, usually measured in square meters per gram, or m^2/gm. (Often this is called simply *surface area,* without the word "specific.")

The main advantage of a high surface area is that, when the powder is fired, it will easily sinter and become a single, large solid body. Coarse powders such as sand do not sinter as well as finely powdered, high area materials such as clay.

The main disadvantage of the large specific surface area is the tendency to *agglomerate,* or form uncontrolled lumps and clusters. Usually the higher the surface area, the more tendency to be agglomerated, other factors being equal.

Agglomerates contain nearly random-size pores, and the largest of these pores can seriously interfere with sintering, as well as causing other problems. In addition, high surface area can raise the viscosity of the slip too much, requiring more liquid such as water to bring the viscosity down to a workable level. More liquid in a given volume means that less powder can be in that volume; in other words, the *solids loading* is less, which can cause a number of problems in later steps. (Each of these points will be discussed in more detail in the chapters to follow.)

The Slip. The first processing step is to mix a liquid with the powder in order to make a flowable or fairly fluid system which can easily be shaped. Another purpose of this step is to break the agglomerates down into individual *particles,* which then are *dispersed,* or evenly spread throughout a small volume of liquid, forming the fluid slip. (The words *dispersion, deagglomeration,* and *deflocculation* have essentially the same meanings from the standpoint of ceramic processing, although there are some subtle differences when these words are used in other technologies.) This processing step is accomplished by mechanical action, usually with a machine that *shears*

---

and their appropriate solutions. For the sake of brevity, many of these problem/solution examples have been put together in that one large table.

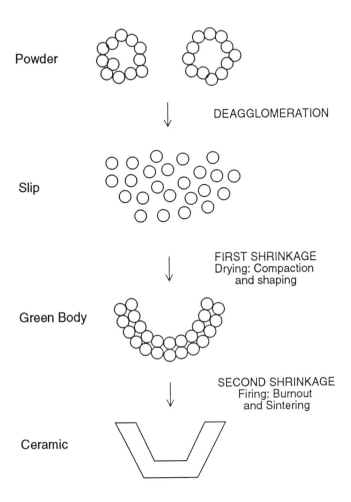

Powder

DEAGGLOMERATION

Slip

FIRST SHRINKAGE
Drying: Compaction
and shaping

Green Body

SECOND SHRINKAGE
Firing: Burnout
and Sintering

Ceramic

Fig. 4.1 Schematic diagram of a ceramic manufacturing process.

the mixture of powder and liquid, examples[1d] being a ball mill, or a mix muller, or an attritor, or a three-roll mill, or a high-shear blade mixer. (The scientific word[1d] for the breaking down of either large agglomerates or large particles into smaller, individual particles is "comminution," which essentially means to make things more "minute," in other words, smaller.) It is important to realize that the breakdown of agglomerates is usually somewhat imperfect. In fact, it is a dynamic process, in which *reagglomeration* is constantly taking place, in a race with deagglomeration. Modern ceramic processing often involves the use of a chemical additive which tends to prevent the process step from going backwards, as much as possible, and this additive is the dispersant. (As mentioned previously, this additive is sometimes called a deflocculant, or a surfactant, but the word "deagglomerant" is never used. However, the words "deagglomerate" and "deagglomeration" are used in this technology.)

When the agglomerates are destroyed, the large pores are also destroyed. In a later step the particles are brought back together without these large pores, although smaller pores will be present at corners where three or four particles come together. These smaller pores are the ones that disappear during sintering, thus raising the density. (The larger pores can not be made to disappear during sintering,[1a,b,c] and therefore they present a considerable problem whenever they can not be destroyed at an earlier step in the process.)

It is often the case that either optimization of the milling or the use of a more powerful dispersant can make dramatic improvements, leading to fewer of the large pores and allowing the use of higher solids loadings without excessive viscosity. However, the reader should take special note of the important statement made above that deagglomeration is driven forward by the mechanical action, and the additive only acts to slow down the reverse reaction.

The First Shrinkage. The next process step in the schematic description of Fig. 4.1 is the first *shrinkage,* in which the liquid is removed. In the manufacture of some products such as *slip cast* dishes and cups, the surface tension of the liquid pulls the powder particles together, and as it slowly dries, the total volume decreases. This causes the shrinkage action. Another word for this shrinkage of overall volume is *compaction.* In the manufacture of complex shapes such as electronic inductor cores, the drying and compaction are separated into two different substeps, *spray drying* and then *dry pressing*. Nowadays this is also being done more and more with simple shapes such as square floor tiles, in order to achieve improved reproducibility, especially as highly automated processes become more common.

If the solids loading (expressed as a volume percentage) in the previous slip is high, then there is less liquid present, and less liquid means there is less shrinkage needed to become the dry green body. If there is less shrinkage, then there is less *variation* in the shrinkage. Less variation leads to less tendency to *crack* during drying, and also less tendency to make a *warped*, distorted green shape. Cracking and warping are two of the main causes of unusable product during manufacturing. The use of a very effective dispersant provides higher solids loading, which tends to minimize these two problems.

Shaping the Green Body. The ancient art of ceramics involved the fluidization of clay by mixing it with a small amount of water. The high viscosity mixture was shaped by hand, using plastic flow. Although this shaping method is still used by artists, it is too expensive for low cost mass production.

Modern methods for shaping are listed in Table 4.1, with simplified diagrams of the equipment shown in Fig. 4.2. Each one is well suited to a certain category of shapes. Thus, the machine operated *extrusion* process is excellent for low cost production of rods or tubes, where the cross section is constant throughout the body. The viscosity must be high enough for the freshly extruded body to be essentially self-supporting. *Roll compaction* is a variant of extrusion, in which the ceramic powder plus a small amount of water is passed between metal rollers, and there is very little friction against the rollers, since their surfaces move along with the powder. However, the wet powder tends to stick to the rollers at random spots, causing "pull-out" pits to be present at the surface of the freshly cast sheet. For higher quality or thinner sheets, *tape casting* makes use of lower viscosity slip, so that any pits will be immediately filled in by flowing slip, before drying takes place. Since tape casting uses a lower viscosity, more solvent is needed, and therefore a long drying chamber must be provided. In tape casting, compaction is not externally driven, but instead it arises from internal forces within the "gelled" binder as it shrinks during drying. Surprisingly, these forces can be just as effective as the large external compaction forces used in the other methods. Table 4.1 is arranged in the order of decreasing viscosity.

The traditional method of hand shaping can be thought of as *pressure casting* that is done with a "mold" (in this case, human hands) that is not porous, where the body is later dried by evaporation. In modern pressure casting (sometimes called *pressure filtration*), the fluidizing water is forced out through a porous mold which absorbs the water by "wicking." The highest green densities can be obtained by

Table 4.1(a)   Modern Ceramic Shaping Methods

Process		Viscosity		Solidification	
	Shapes		Fluidization		Compaction
Dry Pressing	Mostly-Uniform Cross Section	Very. High	None	Plastic Flow of Binder	Yes
Injection Molding	Almost Any	High	Melted Binder	Cooling	No
Extrusion	Uniform Cross Section	High	Water	Plastic Flow	No
Roll Compaction	Thin Sheets	High	Water	Plastic Flow	Yes

Table 4.1(b)   Modern Ceramic Shaping Methods

Process	Shapes	Viscosity	Fluidization	Solidification	Compaction
Tape Casting	Thin Sheets	Medium	Organic or Water	Evaporation	No †
Pressure Casting	Various	Medium	Water	Wicking and/or Plastic Flow	No
Slip Casting	Various	Low	Water	Wicking	No

For typical compositions and problem-solving chart, see Chapter 10.

† No externally applied compressive force, but there is an internally applied tension from the binder.

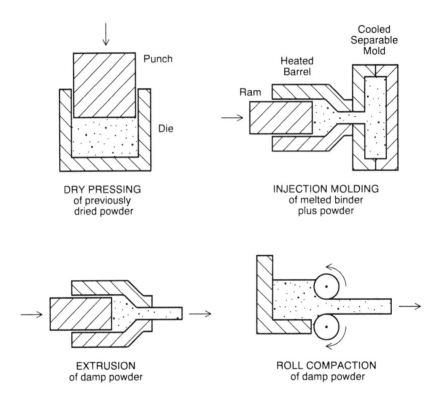

Fig. 4.2a Equipment for shaping. Dry pressing and injection molding are also used in powder metallurgy. (These highly simplified illustrations are not at all drawn to scale. For example, in the molds, the sizes of the micropores are very much exaggerated. See Chapter 10 for some further details.)

this method, because the water is a very effective lubricant which allows the particles to slide against each other while moving to close-packed positions. However, the equipment is expensive.

Slip casting is the traditional method for shaping dishes, sinks, etc., and has been used for hundreds of years with clay based compositions. A porous mold absorbs the water. After the body becomes dry enough to have some self-supporting strength, the mold halves are *separated* for removal.

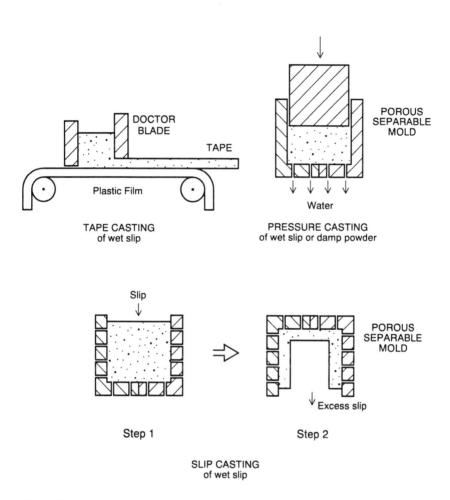

Fig. 4.2b Equipment for shaping. Tape casting is also used in paint and printing ink technology, especially for testing purposes.

By the use of high pressure, viscous dispersions of ceramic powder can be forced into intricate forms with *injection molding*. There is no solvent present to fluidize the powder, and melted binder is used instead. More binder is needed here than with the other shaping methods. This large amount of organic material must then be "burned out" (actually mostly evaporated at high temperature) during the first substep of *firing*, which is done very slowly to minimize warpage.

Even higher pressure is used in *dry pressing*, which requires a binder that can be made to exhibit plastic flow. Because of the extremes in pressure needed to force a solid material into the required shape, the mold (called a "die") is not the splittable type, and so the body must be ejected out of one end. For this reason the body has to have a nearly constant cross section, although some types of step-shaped designs can be made with the use of partially movable dies.

More comparisons of these processes will be covered in later chapters, along with optimized formulations which make use of various organic additives such as binders and lubricants. Also, further explanations of these shaping methods are available in the books by Reed[1b] and Richerson.[1c] After the first shrinkage, the ceramic body is in the green stage, shaped and compacted, and ready to be fired. Generally speaking, but with some exceptions, the fewer large pores the better, at this stage, because of the fact that large pores do not disappear during sintering. In other words, the *green density* should usually be high.

Sintering. The next step is shrinkage number two, which is accomplished by firing. The high temperature, 700°C or much higher, first removes any organic additives that are present ("burnout"), and then it causes sintering. (Sometimes these are done in two separate substeps with separate atmospheres.) Another word for the shrinkage that takes place here is *densification.* If there is a high green density, then less of the *second shrinkage* is needed to provide the final fired density. Thus there would be less warping and cracking, similar to the situation during the first shrinkage. The use of very effective dispersants provides higher green density,* which tends to minimize these two problems during the firing, just as in the case of higher solids loading.

A major driving force that causes sintering is reduction in surface area. A simplified, schematic illustration of this is shown in Fig. 4.3,

---

* To be more precise, the term "packing factor" should be used here instead of green density. However, the two terms are closely related, and a more detailed description will be postponed until other concepts are introduced. Suffice it to say at this point that the green density is actually what is measured directly in most mass production situations, and it is related to the firing shrinkage.

drawn as if the powder particles had cubic shapes. (Actually, some magnetic "ferrite" ceramic powder particles do have cubic shapes,

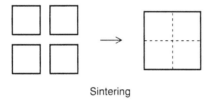

Sintering

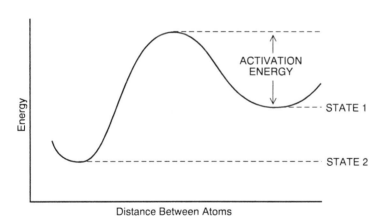

Fig. 4.3 Top figure shows the change in surface energy, which is a major driving force in sintering (see also Fig. 6.14 in Chapter 6). Bottom figure illustrates the theory of "activation energy."

when they are made by hydrothermal coprecipitation, although that is not a common particle shape for other ceramic powders.) Since each cubic particle has six sides, the area of each is six times the area of the small square shown in the diagram. If four of the particles join together and become a single, larger particle, as shown, then the four surfaces indicated by the dashed lines will disappear.

The surface of a solid material almost always consists of stretched chemical bonds, which are in a high energy condition analogous to stretched springs. This energy can be given up as heat, when they become ordinary, unstretched internal bonds that go to the next atoms in the material. Because of the laws of nature, if a stretched spring (or a stretched chemical bond) can relax, it will do so, giving up its energy in the form of heat or some other form.[†] In the energy diagram shown in Fig.4.3, the stretched condition is labeled as State 1, and the relaxed condition as State 2.

In real powder systems, if the four particles are brought together and allowed to touch each other, they will not spontaneously join together. Instead, there is usually a small energy barrier that must be overcome first. For example, water vapor from the air is attached strongly to most ceramic powder surfaces, and this must first be removed in order for the particles to really touch each other with no water in between. Some small amount of energy must be put into the system in order to remove this water.[2] The *activation energy* shown in the diagram is what is required to remove the thin water film, break existing stretched bonds, and do whatever else is required to activate the surfaces and allow the particles to coalesce into a single, larger one.

The above sequence, although somewhat simplified, illustrates the facts that are observed experimentally: (1) in order to sinter, particles must be brought into contact; (2) heat must be provided to supply a certain amount of activation energy; (3) if conditions are right, the

---

[†] The spreading of thermal energy can be monitored in terms of increasing "entropy." In teaching thermodynamics, the author has found it to be convenient to express entropy as "the volume in which the heat is contained." Just as the volume that encloses gas molecules tends to increase with pressure, the volume that encloses heat (that is, the entropy) tends to increase with temperature, and the mathematical relationships are quite analogous except for a sign difference.

[2] J. C. Chang, F. F. Lange, and D. S. Pearson, J. Am. Ceram. Soc., 77 (1994) 19; J. N. Israelachvili, et al., J. Am. Ceram. Soc., 77 (1994) 437.

particles will then sinter into a single, pore-free mass of material, and a small amount of *free energy* (the difference in the energies of states 1 and 2 on page 101) will be given off by the system, usually as heat.

Following the ideas of the above discussion, the higher the surface area per gram of powder, the more surface energy is available as a driving force for sintering. Therefore, a measurement of the specific surface area of the starting powder should tell the ceramist how likely it is that the powder will sinter, if it is tightly compacted together and heated.

It should be noted that there are other ways to think about this particular driving force, which are discussed in classical treatments of ceramics theory, in terms of grain boundary motion, contact angles, pore shapes, and so on.[1a] Also, there are other possible driving forces and many possible characteristics of the powder that might affect sinterability, in addition to the surface area. Two examples taken from the author's research are the strain energy involved in dislocation defects within the particles[3] and impurities that can enhance bulk diffusion,[4] but there are many more discussed in the literature of sintering.[1a] It is the author's experience that the one powder characteristic that is most important for predicting sinterability, and also for determining the need for organic chemical additives, is the specific surface area. This is conveniently measured by the B.E.T. method,[5] named after the initials of the three authors who perfected it. (It might interest the reader to know that E. Teller is the same scientist who played a major role in the development of the hydrogen bomb.) Other powder characteristics are occasionally important in special cases, but the B.E.T. value has great significance in practical manufacturing situations.

Aside from the driving force, the actual pathway or *mechanism* for sintering which is most commonly encountered in industrial ceramic processes is *liquid phase sintering*.[6] Except for a highly specialized

---

[3] D. J. Shanefield, et al., in "Ceramic Processing Before Firing," G. Y. Onoda and L. L. Hench, Eds., J. Wiley, New York (1978) 141.

[4] W. van Rijswijk and D. J. Shanefield, J. Am. Ceram. Soc., 73 (1990) 14.

[5] S. Brunauer, P. H. Emmett, and E. Teller, J. Am. Chem. Soc., 60 (1938) 309; S. Lowell, "Powder Surface Area and Porosity," Chapman & Hall, N.Y. (1991).

[6] R. M. German, "Liquid Phase Sintering," Plenum Press, New York (1985).

process for firing silicon carbide entirely by solid state diffusion,[7] and except for hot pressing, it is the author's opinion that almost all commercially viable ceramics manufacturing processes make use of liquid phase sintering. (The reader who is particularly interested in this point might turn briefly to the section on *liquid phase sintering* in Appendix I at the end of the book.) In spite of the fact that this appears to be generally true, little or no attention is devoted to liquid phase sintering in most ceramics textbooks or in most courses of study.

In the liquid phase sintering mechanism, a small amount of impurities or *inorganic* additives melt at the firing temperature. (Following the terminology used in powder metallurgy, *impurities* are minor constituents that are not added purposely by the engineer, while *additives* are purposely included. In this case, it is *inorganic additives* that are involved, not organic, so that the desired material will remain after firing and not burn away. However, "organometallic" additives are sometimes used, where part of the molecule burns away but the other part remains to aid in sintering.)

Some of the ceramic powder (or metal, in powder metallurgy) dissolves in the melt and then recrystallizes on the surfaces of larger, undissolved powder particles, making them larger still. The growing *grains* coalesce into one continuous ceramic body. During liquid phase sintering, surface energy reduction is still the most significant driving force. (This is proportional to the surface area reduction, so the specific surface area is still a useful thing to measure, even with the liquid intermediate step.) The free energy change during sintering is approximately the same, whether liquid or solid states are involved, and only the activation energy is lowered by the liquid pathway. These materials which enhance thermal densification are referred to as *sintering aids.*

In some cases the liquid is mostly squeezed out from between the growing grains by a de-wetting action that takes place as the new grain surfaces become relatively low energy crystal faces.[8] Toward the end of the densification step, when most of the liquid has been squeezed out

---

[7] J. A. Coppola, et al., U.S. Patent 4,312,954 (1982); R. Hamminger, et al., J. Mat'ls. Sci., 18 (1983) 3154.

[8](a) C. A. Powell and A. H. Heuer, J. Am. Ceram. Soc., 73 (1990) 3670 [see Fig. 5]; (b) H. Yan, W. R. Cannon, and D. J. Shanefield, J. Am. Ceram. Soc., 76 (1993) 166 [see Fig.6].

and it no longer forms a continuous network, strictly solid state diffusion might become the dominant mechanism for the remainder of the sintering step.

The main reason why liquid phase sintering appears to be so common in industrial ceramic sintering is that many combinations of impurities can form eutectics which melt at fairly low temperatures, and as little as about 1% of this liquid phase can be sufficient to completely coat a typical ceramic powder with a layer of liquid which is just one molecule thick (a "monolayer"). Figure 4.4 illustrates a sample calculation for this kind of *monolayer*, in approximate terms. Thus, even 99% pure alumina, for example, contains enough impurities to coat most of the ceramic grains. (Figure 4.4 also contains information about the approximate amount of organic additive such as a *dispersant* or a *lubricant* that must be used to provide a monolayer or adsorbed material: it is often roughly 1%. More about this will appear in later chapters.)

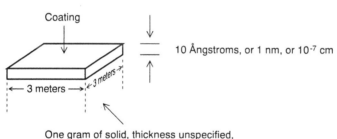

Coating

10 Ångstroms, or 1 nm, or 10^{-7} cm

3 meters

3 meters

One gram of solid, thickness unspecified, sides and bottom not coated, specific surface area of coated part = 10 m²/gm.

Volume of adsorbate = (3.3 x10² cm) (3.3 x 10²cm) (10^{-7} cm) = 10^{-2} cm³

If the density of the adsorbate is about 1 gm/cc, then the weight is about 10^{-2}gm.

On 1 gm of solid, this is approximately 1%.

Fig. 4.4 Estimate of the weight percentage that a monomolecular coating comprises on a powder having a specific surface area of 10 square meters per gram.

Another general principle which favors liquid phase sintering is the fact that most combinations of oxides can form eutectic mixtures when fused, having melting points below those of the pure materials. Also, at the high temperatures involved in sintering, there is at least some degree of solubility of most ceramics in mixed oxide eutectics. These are all reasons why liquid phase sintering appears to be quite common.

As one particularly illustrative example, it was shown during an extensive study at AT&T that 99.5% aluminum oxide ("alumina") sinters via a liquid phase, in spite of the fact that it is considered to be a relatively "high purity" material. The 0.5% impurities were mostly calcia, magnesia, and silica, with a eutectic melting point of only about 1300°C, and the firing was done at 1500°C. At least one monolayer of glassy impurity was found at the grain boundaries after firing.[9] Most of that impurity mixture had been squeezed out of the body and onto the outer surface, appearing as a skin, which was detected there by sputtered ion mass spectroscopy[10] While it has not been directly proven, it can be surmised that most other materials that have at least 0.5% impurities probably sinter initially in the same manner, since they usually contain more eutectic-forming impurities than this. (Note: solid phase sintering might still be available for the last bit of densification, after the liquid is squeezed out.)

The presence of the monolayer or so of impurity at the grain boundary does not necessarily degrade the desirable properties of the ceramic. For example, the thermal conductivity of liquid phase sintered aluminum nitride can be 165 W/Km,[8b] and the strength of liquid phase sintered silicon carbide[11a] can remain quite high at 1400°C,[11b] in spite of a small amount of sintering aid remaining in the fired body. In some cases the grain boundaries might be remarkably clean and free of the liquid additive after the de-wetting, as shown by high resolution electron microscopy.[11b,12] However, some contrary

---

[9] D. J. Shanefield, Matls. Res. Soc. Proc., 40 (1985) 69 (see particularly page 76).

[10] R. E. Mistler, P. T. Morzenti, and D. J. Shanefield, Am. Ceram. Soc. Bul., 53 (1974) 564.

[11](a) L. Cordrey, D. J. Shanefield, and D. E. Niesz, Ceram. Trans. 7 (1989) 618; (b) X. Chen, M.S. Thesis, Rutgers University (1990); (c) X. Chen, D. J. Shanefield, D. E. Niesz, and Eva Koh, Am.Ceram. Soc. Bul., 69 (1990) 496, further reports to be published.

[12] S. M. Wiederhorn, et al., J. Am. Ceram. Soc., 77 (1994) 444.

evidence obtained by an "unfocused" microscopy technique indicates that a monolayer of amorphous additive might still be there.[13a]

An interesting trend in advanced ceramics involves the use of *organometallic* compounds as a means to put sintering aids into the composition.[13b, 11b,c] (Some chemists insist that these compounds should logically be referred to as "metalorganic," but the first terminology is more commonly used, logically or not.) These are partly organic and also partly inorganic, with an example being "aluminum tri-sec-butoxide" (or "ATSB"), which can be used as a sintering aid for silicon carbide instead of using powdered aluminum oxide. ATSB is a liquid, and therefore it can be mixed into the composition in a uniform manner, being divided down into individual molecules which spread out evenly around each ceramic powder particle. The composition is then exposed to air, then fired, and the ATSB then becomes aluminum oxide of extremely high surface area. (In this case the organometallic additive is said to be a *precursor* of the alumina.) The two factors of being evenly spread around the particles and having an ultrahigh surface area can cause the liquid sintering aid to be very effective, compared to the addition of a powdered solid sintering aid. Sintering times and temperatures can both be decreased by the use of these liquid materials, and grain growth can thus be less than with the use of an powdered sintering aid, yielding better properties in the fired ceramic product. For example, silicon carbide can be sintered at 1850°C by this means,[11b] while the addition of powered sintering aids would ordinarily require about 2150°C and would result in more grain growth. Organometallics are also being used on a small scale as precursors of the dispersant, of the binder, and of the main bulk ceramic, although a number of problems such as excessive outgassing are still being worked out.

The Fired Ceramic Product. Generally, sintered ceramic bodies have rather limited strengths, compared to metals, plastics, and even wood, because of the tendency toward brittleness, which is the most serious disadvantage of ceramics. However, the great advantages of

---

13(a) L. K. L. Falk, Am. Ceram. Soc. Bul., 73 (1994) 259, further reports to be published. (b) This technique, used for a silicone additive, was reported by the author 20 years ago, in the first paper cited in reference 18 (see particularly page 418 of that paper), but it has taken a long time to become a generally-used additive method, even though it is very effective for the addition of sintering aids.

resistance to corrosion, their high temperature capabilities, and their hardness are sufficient to make ceramics extremely useful and even indispensable for such large scale uses as sinks, dishes, tiles, crucibles, cutting tools, and turbine blades. In addition, they have many unique properties that are useful for low cost but high performance electronic devices, such as hermetic packages, piezoelectric sensors, and high voltage insulators.

One disadvantage is a tendency for fired ceramics to be porous, although this is not always true. A material having *continuous porosity*, where the pores are interconnected like those of a sponge or a filter, has several serious disadvantages for practical uses. A cup made of such material could not, of course, hold water without it draining slowly out through the pores. Even if the drainage was extremely slow, the cup would absorb food, etc. from previous uses and be unsanitary for use as dinnerware. Such ceramic could not be used for most electrical purposes as insulators, etc., because of uncontrolled absorption of water vapor from the atmosphere. Also, highly porous materials tend to be mechanically weaker than fully dense material.

If the porosity is limited to about 3% or less, it is usually *closed porosity* and is not continuous. This is satisfactory for many purposes, and almost all practical ceramics do have some porosity of one kind or the other. Two other ways to compensate for the great tendency of ceramic materials to be at least somewhat porous are to cover the material with a thin layer of glass called "glaze," or to permeate the whole body with a "vitrification" glass. At any rate, porosity in the fired ceramic is something that needs to be either minimized or at least carefully controlled.

Even a small number of very large pores can have a significant weakening effect in ceramics. This is because the cracks that can cause failure usually start at unusually large pores or at surface scratches. Once a crack starts in a brittle material, it can propagate further with relatively little applied force. Since large pores are sometimes caused by agglomerates that are not properly broken down during processing, a detailed study of these entities is worthwhile.

4.2 Agglomerates

The Causes of Agglomeration. Dry agglomeration is caused mainly by two things, van der Waals forces[14] and H-bonds.[15] The terminology in widespread use in the ASTM specifications, etc., uses the word "particles" for those small objects which are weakly stuck together in lumpy form but which are easily separable by sieving, milling, ultrasonic action ("ultrasonicating"), and other moderately strong stirring ("agitation") methods.

The lumps themselves are called "agglomerates." If the small *particles* have somehow become stuck together very strongly, so that they can not easily be separated, they are called "aggregates," or "hard agglomerates." Exactly how strongly they are held together, and exactly how easily they can be separated have not been standardized throughout the industry in terms of numerical values or tests. The strengths of these interparticle bonds that lead to agglomeration can be estimated, at least in a comparative or relative sense, by their resistance to being broken down in wet versus dry vibrating sieves.[16] A practical standard for use within a given organization such as a corporate quality control laboratory is to progressively increase the deagglomeration action (ultrasonic energy or time, etc.) in a carefully documented manner, until little further effect is seen. An example of this type of "plateau effect" is to ultrasonically disperse a previously agglomerated powder at 40 kHz and 1 watt in 1% aqueous sodium hexametaphosphate solution for 1 minute, 3 min., 10 min., 30 min., and 100 min. and measure the apparent particle sizes after each treatment. The apparent size (or possibly the median size if there is a wide range) would be expected to decrease from about 1 to 10 minutes, but possibly there would be little further decrease between 10 and 100 minutes, as in Fig. 4.5. If so, then the size measured after 10 minutes would arbitrarily be called the "particle size." This would also be the "aggregate size." Possibly some other technique could determine whether or not

---

[14] H. Rumpf, Chem.-Ing. Tech., 30 (1958) 144 (in German).

[15] D. J. Shanefield, Matls. Res. Soc. Proc., 40 (1985) 69; R. G. Horn, J. Am. Ceram. Soc., 73 (1990) 1117 [see page 1128, water effects].

[16] D. W. Johnson, Jr., et al., Am. Ceram. Soc. Bul., 51 (1972) 896.

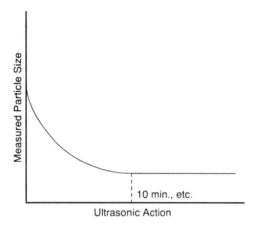

Fig. 4.5 "Particle" size measurements after various ultrasonic times, energy settings, etc.

each *particle* is actually made up from several smaller particles strongly held together. (To some degree this can be done with a "pycnometer," to determine whether pores exist inside the particle, and x-ray line broadening can determine whether each particle is made up from more than one crystal.[2]) However, in the author's experience, further attempts are not of any practical use. Instead, considering the particle and the aggregate to be the same thing is a useful approximation. Making up a standard dispersion procedure and deagglomerating until a plateau is reached is a useful way to define the particle, instead of attempting to find some absolute standard.

Types of Agglomerates. One type of agglomeration involves the lumping together of groups of particles, with large pores inside each one. As was mentioned earlier, it is well known in ceramic technology that pores above a certain size in the green body will fail to shrink during sintering.[1a] The exact value of this critical size varies according to many conditions of temperature, purity, etc. There are many reports of experimental studies that trace the passage of a few agglomerates and their pores through the whole process, when the milling and the dispersant have been inadequate, and the result has usually been poor strength or other serious defects.[17]   This is one of the reasons why effective deagglomeration is important.

Another type of agglomeration is not a condition of local lumpiness but is instead a kind of widespread, diffuse, and relatively weak sort of bonding between all of the particles in the slip. The entire body of slip might be thought of as a single, weak agglomerate. This makes itself known as very high viscosity, or a very "thick" slip, which will not flow easily, and the viscosity can be too high for practical processing. Indeed, the viscosity is often used as a measure of residual agglomeration.   Various experimental studies have reported that ineffective dispersants or insufficient milling can lead to excessive viscosities because of this widespread type of agglomeration.[18] Contrariwise, improved dispersant or milling lowers the viscosity.

Agglomerates can consist of a number of randomly-sized units, ranging from small to large. If this is the case, and the dispersion and milling are poor, the resulting green body might have a high average porosity. The overall green density would therefore be low, since much of the volume would consist of vapor-filed pores instead of highly dense ceramic. Many reports of experimental studies have shown that poor deagglomeration leads to low green density.[19]

The overall green density is used as a routine quality control tool to indicate the degree to which the powder particles are in close enough contact to sinter properly. If the green density is too low, the body might not sinter to a high enough fired density. Experimental studies

---

[17] D. E. Niesz and R. B. Bennett, p. 61 in G. Y. Onoda and L. L. Hench, "Ceramic Processing Before Firing," J. Wiley, New York (1978); I. Aksay, Advances in Ceram., 9 (1984) 94; A. G. Evans, et al., J. Am. Ceram. Soc., 67 (1984) 741; F. F. Lange, J. Am. Ceram. Soc., 72 (1989) 3; L.-Y. Chao, et al., J. Am. Ceram. Soc., 75 (1992) 2116.

[18] D. J. Shanefield and R. E. Mistler, Am. Ceram. Soc. Bul., 53 (1974) 416; D. J. Shaw, "Introduction to Colloid Chemistry," Butterworths, N.Y. (1980) p.208; B. M. Moudgil, et al., Ceram. Trans., 26 (1992) 66; I. Tsao and R. A. Haber, Ceram. Trans., 26 (1992) 73.

[19] D. W. Johnson, Jr., et al., Am. Ceram. Soc. Bul., 51 (1972) 896; J. W. Halloran, Advances in Ceram., 9 (1984) 67; A. Kerkar, et al., J. Am. Ceram. Soc., 73 (1990) 2879; K. Nagata, Ceram. Trans., 22 (1991) 335 (see Fig. 3); K. Nagata, Ceram. Trans., 26 (1991) 205 (see Fig. 1).

have repeatedly shown that low green density can lead to low fired density, other factors being kept equal.[20]

There can be additional problems caused by low green density. One is that the *second shrinkage* (during firing, as indicated in Fig. 4.1) will have to be too great in order to achieve a useful final density.   A greater amount of firing shrinkage can lead to more random variation in the shrinkage itself, and that leads to more cracking and warping, just as it does in the first shrinkage. Even without warping or cracking, it can lead to more variation in the final size, so that more of the product is "out of specification" regarding size.

In specialized aspects of ceramic processing such as high speed slip casting, a type of agglomerate called a "floc" is actually desirable. In addition, large, round agglomerates are sometimes desirable as a feed material for dry pressing. These examples will be discussed in later sections.

4.3 Optimum Surface Area

The specific surface area of a ceramic powder is generally proportional to the surface energy. In Fig. 4.6, a rough diagrammatic relationship between these is shown by the bottom axis, the left-hand axis, and the diagonal straight line. Since the surface energy is a major driving force for sintering, the fired density might be expected to also depend on the surface area. Although that is true, as mentioned above, another factor also comes into play, and that is the fact that powders with high specific surface areas tend to pack together poorly because of agglomeration, and the achievable green densities are therefore low. This relationship is also shown, again only in a rough diagrammatic manner, by the diagonal curved line. The very low green densities that can be achieved with high surface area powders do not provide sufficient interparticle contact for effective sintering. At any given value of surface area, the heights of both the green density line and the

---

[20] I. B. Cutler, p. 21 in G. Y. Onoda and L. L. Hench, "Ceramic Processing Before Firing," J. Wiley, New York (1978); I. A. Aksay, Ceram. Trans., 1B (1988) 663; J. S. Reed, et al., Amer. Ceram. Soc. Bul., 71 (1992) 1410.

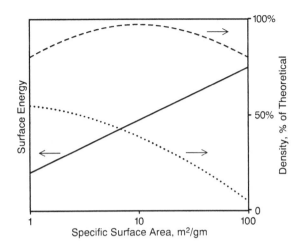

Fig. 4.6 Typical effects of surface area on green density (dotted curve), interacting with surface energy (solid line), to produce fired density (dashed curve).

surface energy line can be added, to yield a rough estimate of sinterability. Therefore, with very high surface area powders, the final fired densities will be low, as shown by the topmost curve.

At present, the optimum value of specific surface area from the standpoint of sinterability appears to be at very roughly 10 square meters per gram. For powders that are easy to sinter, such as barium titanate,* the optimum shifts to about 5 m^2/gm, because the high driving force is not necessary, but obtaining the highest possible green density is desirable in order to have only a small variation in firing shrinkage. For powders that are difficult to sinter, such as silicon carbide,† the optimum shifts to about 15 m^2/gm, because of the need to have as much driving force as possible, but without sacrificing too much reproducibility in shrinkage. These values are based on current experiments with the best available dispersants, milling, compaction, etc.

---

* Sinterable at about 1,000°C.

† Sinterable at about 2,000°C.

in seeking better dispersants. Better dispersants might lead to both a lower temperature for the sintering of difficult materials, and also to better control over shrinkage. In addition, in cases where improved dispersants have enabled the use of finer powders, the fired grain size has been smaller, leading to improvements in smoothness,[21] fracture toughness,[11b] and other properties.

Chapter 10 lists typical formulations for use with each shaping process, along with some other processing details. Explanations of the theoretical principles underlying these particular formulations appear in the following chapters.

---

[21] D. J. Shanefield and R. E. Mistler, Am. Ceram. Soc. Bul., 53 (1974) 416, 564.

‡ It is a challenge to ceramists that the same type of covalent chemical bonding which provides us with high temperature strength and high hardness also makes the material *difficult to sinter*, mostly because of low diffusion rates. This is the reason why we need the very fine (high surface energy) powders, and the next logical step is that we then need temporary deactivation (for processing), which is exactly what the dispersant provides.

Powder metallurgists have an easier task, since metals are (1) ionically bonded and therefore have (2) high diffusion rates, and thus are (3) *easy to sinter*. This allows the use of (4) coarser powders, which (5) pack better than typical fine ceramic powders (also the metal particles are deformable), and therefore (6) the green bodies shrink much less during sintering.

5. PARTICLE CHARACTERISTICS

5.1 The Packing of Powders

The Sizes Needed for Closest Packing. Because of the fact that close contact between powder particles is necessary for effective sintering, it is important to optimize the way that the particles pack together in the green body. If the particles were spheres, and they were arranged in the "closest packing" that is geometrically possible, they would occupy 74% of the volume, with empty spaces (small "pores") occupying the rest of the volume. An illustration of just three of these spheres is at the top of Fig. 5.1. The percentage of the volume that is occupied by the solid spheres is listed at the upper left corner of Table 5.1. In real ceramic green bodies, the particles are rarely spherical, although they are often spheres that are somewhat distorted. Also, they are very rarely in the *closest packed* configuration, although sometimes they can be nearly closest packed, and it is instructive to compare the ideal case to these.

In three dimensions, there are two different ways that closest packing can be achieved, cubic and hexagonal, depending on the arrangement of the next layer (which is off the plane of the paper in Fig. 5.1 but still parallel to that plane). For the purposes of this discussion any differences between the cubic and hexagonal configurations will be ignored, because they each have the same* *packing factor, 74%.*

---

* The term *packing factor* means the same thing as *packing fraction, packing efficiency, packing density, volume fraction of solids,* and *solid occupancy,* except that "packing fraction" is expressed as a decimal fraction like 0.74, instead of being a percent like 74%, which is sometimes used for the other terms. If there is no other material such as the binder present, then *percentage of theoretical density* also means the same thing, and this is usually proportional to the *green density.* However, the presence of some binder would make the latter two terms mean something slightly different.

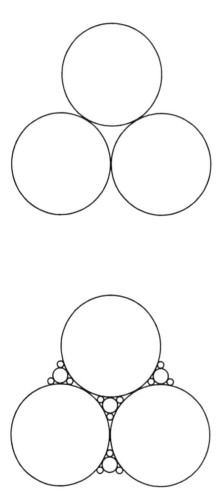

Fig. 5.1 Closest packing of monosize and trisize particles in two dimensions.

Table 5.1   Particle Packing Factors

Sizes	Closest Packed	Lubricated and Pressed	Shaken and Settled
One size (spheres) *Large*	74%	~ 65%	~ 60%
Many sizes, (and shapes) *Coarse Powder*	> 95%	~ 90%	~ 80%
Many sizes, (and shapes) *Fine Powder*	——	~ 55%	~ 33%

Although there are only three spheres shown in the figure, a larger number of particles could also be arranged in a closest packing configuration. If these were real metal ball bearings, and they were shaken and then allowed to settle by gravity, they would then have nearly-random defects in their structural arrangement, and the packing would no longer be the closest possible. On the average, if this experiment is repeated many times, the packing factor is approximately 60%, as indicated at the upper right of Table 5.1. (The snaky "tilde" mark means "approximately.") If an assembly of ball bearings is lubricated with grease and then pressed together in a cylindrical mold of some sort, a packing factor of roughly 65% can be obtained.

The lower half of Fig. 5.1 shows that three sizes of particles can lead to a higher packing factor, because small particles fit into the *interstices* (small pores) between the larger ones, and then particles that are even smaller yet can be fitted into the remaining interstices. The use of more different sizes could result in higher packing factors. In three dimensions, rather than the two dimensional closest packing shown in the figure, the optimal sizes would be slightly different, since the smaller particles would be sitting up off the plane of the paper.

Many studies of closest packing have been reported.[1]   In two dimensions, both theoretical and experimental work has been done, and the agreement is excellent.   However, in three dimensions, it is extremely difficult to calculate the closest packing particle size *distribution* (that is, the number of particles of each particular size), and the reported work has been largely experimental.   The author has experimented with spheres of eight sizes, using croquet balls, cork fishing floats, glass marbles, vitamin pills, and lead shot, and about 95% packing was achieved.   The size distribution is shown numerically in Table 5.2 and graphically at the top of Fig. 5.2.   Most of the volume of material is in only a few large spheres, with very little volume in the many medium size and small size particles.

Table 5.2.   Sizes of Approximately Closest Packed Spheres

Sphere Diameter, mm	Volume of A Single Sphere, mm^3	Number of Spheres In The Unit Cell	Total Volume,* All Spheres of That Size, mm^3
3	14	48	672
6	113	24	2,712
7	180	36	6,480
12	905	12	10,860
14	1,437	24	34,488
17	2,572	8	20,576
32	17,157	4	68,628
77	239,040	4	956,160

* Per one unit cell of the four largest spheres

[1] G. L. Messing and G. Y. Onoda, J. Am. Ceram. Soc., 61 (1978) 363; G. Y. Onoda, Advances in Ceramics, 21 (1987) 567; J. S. Reed, "Introduction to the Principles of Ceramics Processing," J. Wiley, New York (1995); R. M. German, "Particle Packing Characteristics," Metal Powder Industries Federation, Princeton, NJ (1989).

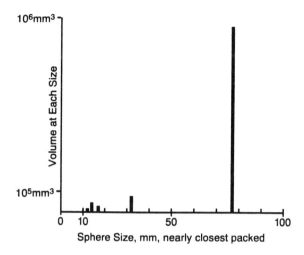

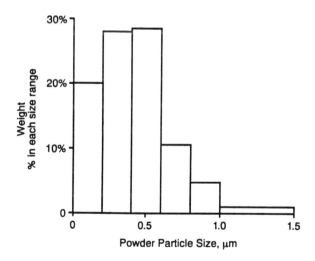

Fig. 5.2 Approximately-closest packed large spheres size distribution (top figure), versus real ceramic powder particle size distribution (bottom figure).

The Sizes of Real Ceramic Powders. A typical powder used commercially to manufacture modern, high performance ceramics has a particle size distribution like that shown by the other bar graph in Fig. 5.2, and it is evident from the comparison that the ratio of sizes in the commercial powder is very different from closest packed. First of all, the powders that are commonly used are whatever can be manufactured at low cost by industrial techniques such as calcining and milling, not necessarily what fits the scientific theories best. Secondly, a higher percentage of *fine* (that is, small size) particles is necessary in order to provide high surface energy (proportional to surface area) and thus promote easy sintering, even though this discourages close packing. Also, a closest packed distribution would not flow readily enough for shaping, since there is not enough empty space for the particles to move between each other easily, even with slight rearrangements, but the commercial powder does flow quite easily.

Even if the commercial powder could be arranged one particle at a time by some kind of a micro-tweezers apparatus, it could not really pack well, since it contains many large pores where none of the relatively small particles exactly fits an available hole. For this typical commercial powder, the lubricated and pressed packing factor is roughly 55%, as indicated in Table 5.1. Thus there is about 42% volumetric shrinkage during sintering if 97 *percent of theoretical density* is to be obtained in the *fired* product,[**] leaving 3% closed, discontinuous porosity.[†] This is a large amount of shrinkage, and therefore further improvements in green body packing factors are constantly being sought, with the exception of cases where final porosity is actually desirable (filters, thermal insulators, etc.).

[**] For use in "advanced" or "high performance" applications, usually a 96% or higher percentage of the theoretical density will mean that there is only a negligible amount of *continuous* or *open* porosity, which could otherwise absorb undesirable water from the environment. When measuring the fired density using the Archimedes method, a water absorption test should be done after boiling the sample in water and then cooling it, to determine that there has not been more than 0.3% water absorbed. (The 0.3% value is the usual random error in this measurement.) Then the Archimedes method will be testing the true *bulk density* (as defined later in Fig. 8.3), rather than the misleading *apparent density,* and the *percent of theoretical density* will be satisfactorily high.

The size distribution of a ceramic powder is measurable by several different techniques,[2] depending on the size range. For coarse powders, vibratory wet sieving is a practical means, because it can continuously deagglomerate the large particles.[3] For very fine powders, the sieves can not be made small enough for the sizes that are involved.* Probably the most commonly used method for fine powders is *sedimentation,* where the powder is allowed to settle after first being deagglomerated in a liquid. The deagglomeration is done ultrasonically with a good dispersant, for longer and longer times until no further reduction in the measured size is noted. Sedimentation instruments are available which graph the size data automatically, provided the powder has been properly deagglomerated.

The Bell Shaped Curve. The amounts of material in each of many *statistical cells*, or small ranges of sizes (like 0.3 to 0.4μm), can be plotted as a bar graph, as shown in Fig. 5.3. Ordinarily, we would expect to see a nearly-symmetrical bell shaped curve ("Gaussian," or "normal" curve), indicated by the solid line in the figure. If each cell is of only "incrementally" small width, a Gaussian normal distribution of sizes can be expressed as

$$y = Ae^{-b(x-c)^2}$$

where y is the height of the bell shaped curve at a given point. That is, y is the "frequency," which tells us how many times any given size, x,

---

[2] T. Allen, "Particle Size Measurement," Chapman & Hall, New York (1990).

[3] D. W. Johnson, Jr., et al., Am. Ceram. Soc. Bul., 51 (1972) 896.

† In this case, the "shrinkage factor" is $0.97^{-1/3} / 0.55^{-1/3} = 0.83$. The fired length (which in this case is $0.97^{-1/3}$) can be predicted by multiplying the green length (which is $0.55^{-1/3}$)by 0.83. The firing shrinkage is $1 - 0.83 = 0.17$, which ceramists would usually express as "17% shrinkage." Sometimes the 0.17 is written as (green length) – (fired length) / (green length). Further data appears later, at the bottom of Table 8.1.

* When sieves are used in the small size ranges, such as 325 mesh, what is measured is most often the agglomerate size, not the actual particle size.

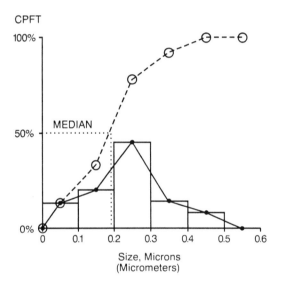

Fig. 5.3 Normal distribution and cumulative normal distribution, on a linear scale.

will occur. The symbols A and b are constants,‡ and e is natural log base, 2.718. The symbol c is the "median" size, in other words, the size at which half of the particles are larger and half are smaller.

‡ The constant A is usually $(2\pi)^{-1/2}$, and b is usually 1/2 in ordinary random distributions.

Particle size distributions of ceramic particles are commonly expressed in terms of *cumulative percentage finer than,* or *CPFT*. By way of explanation, if the height of each bar is added to the total height of all smaller bars, then that would be a curve that grows in height as the sizes get larger. (See the dashed line in Fig 5.3 as an example.) This is a *cumulative* distribution, where at each point on the curve, the total weight of all the smaller (*finer*) particles is shown. These weights can also be written in terms of a *percentage* of the overall total weight. That is the CPFT type of plot. Smaller cells would provide a smoother curve, and at vanishingly small sizes, the curve would be the "integral." The *median* size is the 50% CPFT point. It should be noted that Fig. 5.3 is only a hypothetical example, which was graphed to illustrate the mathematical principles, and it does not represent a real ceramic powder.

The Lognormal Distribution. Because of the methods used for making inexpensive ceramic powders, their size ranges are not usually "normal distributions." The size data for most of the inexpensive ceramic powders can only be made into a CPFT curve that has a fairly smooth, straight-line central portion if the size axis of the graph is logarithmic, instead of being linear. When the size distribution turns out to fit best with this semilog type of bell shaped curve, it is called "lognormal." In other words, if the logarithm of the size is substituted into the above equation instead of the directly measured size, the curve becomes *lognormal.* Since so many ceramic powder distributions are lognormal,[4] semilog paper is often used to make the distribution become nearly a straight line.

Although ceramists do not ordinarily make use of this, the same data would usually be even more linear if it were plotted on *log probability* paper,[5] which stretches out both ends of the vertical scale. A similar linearizing effect can be seen when Weibull paper is used. Either one of these plotting techniques makes it easier to extrapolate beyond the measurable range of sizes, or to see when a new lot of powder is out of specification because of a non-standard size

---

[4] J. S. Reed, et al., Ceram. Trans., 1B (1988) 733.

[5](a) F. H. Steiger, ChemTech, (April 1971) page 225 [see particularly Fig. 2.].
(b) R. M. German, "Particle Packing Characteristics," Metal Powder Industries Federation, Princeton, NJ (1989) page 209 [see particularly Fig. 8.19a].

distribution, which is more immediately and positively apparent when straight lines are compared. If the exponent, 2, is substituted by other numbers, n, then this becomes a special case of the even more general "Weibull distribution."

Additional Porosity. After shaping, real assemblies of fine particles in green ceramic bodies contain large pores because of:

(1) non-ideal size distributions, and

(2) random variations in the particle positions.

The latter is due to the fact that, even if they are being pressed together, as the particles move into their final placement, they are not able to find the positions of best possible packing because of:

(2a) previous agglomeration,

(2b) new agglomeration due to friction, and

(2c) other particles blocking the way.

A good dispersion step (such as wet ball milling with a dispersant) in the overall process will tend to minimize the effect of agglomeration, and a good lubricant (sometimes the same as the dispersant) can minimize friction. The compaction forces used in dry pressing or pressure casting can very much decrease porosity, or at least control it to a reproducible level. Ultrasonic or other vibration (sometimes just periodic tapping) can tend to minimize the blockage of some particles by others.

None of the above pore-reducing expedients can be done with 100% effectiveness. Therefore, even with well dispersed and well lubricated particles that are vibrated during compaction, there is a "hierarchy" of pore sizes,[6] and a mathematically describable *pore size distribution.*[5b] As examples, Table 5.1 shows that a typical fine powder when allowed to settle in the dry state under 1 g of gravity has a packing factor of roughly 33%. (The weight per overall volume of vibrated powder is sometimes called the *tap density.*) As shown in the table, lubrication

---

6 I. A. Aksay, Ceram. Trans., 1B (1988) 663.

and pressing can improve the packing to about 55%. Across a broad range of ceramic materials, although not all of them, it is a rule of thumb that a packing factor of somewhat above 50% is necessary for pressureless sintering to be able to proceed to the point of no measurable "open" or continuous porosity. Of course, these numbers vary considerably according to the materials and the process details.

Modified Particle Size Distributions. Assuming that there is good dispersion, the particle packing can be increased very much by adding a small amount of a coarser powder, according to various mathematical schemes,[7] or by "calcining" (heating at about 900°C to coarsen, usually followed by ball milling[*] to break down *aggregates*). However, if carried to extremes, the sinterability might be decreased by the excessive coarseness. The packing factor can sometimes be increased by adding finer powder, which can usually increase sinterability,[8] but carrying this too far will increase the viscosity of the resulting slip a great deal.[9] With a powder that is quite well dispersed in a fluid, optimization of the particle sizes can often increase the packing without raising the viscosity excessively, if it is done according to certain mathematical relationships between the largest and the smallest particles. One of the more popular size distributions for accomplishing this in practice is the Funk-Dinger[10] distribution:

$$\text{Cum. \% Finer Than} = \left(\frac{s - min}{mx - min} \right)^n$$

where s is a size, min is the smallest size in the distribution, and mx is the maximum size in the distribution. High packing fractions are obtained with $n = 0.37$, but easier flow in slip casting of clays was usually found by Funk and Dinger to occur with $n = 0.27$.

---

[7] G. W. Phelps and M. G. McLaren, page 211 in "Ceramic Processing Before Firing," G. Y. Onoda and L. L. Hench, eds, J. Wiley, New York (1978); G. Y. Onoda, Advances in Ceramics, 21 (1987) 567; J. E. Funk and D. R. Dinger, Am. Ceram. Soc. Bul., 67 (1988) 890; P. A. Smith and R. A. Haber, J. Amer. Ceram. Soc., 75 (1992) 290.

[*] Dry ball milling is often more effective than wet milling for this purpose.

[8] I. A. Aksay, Ceram. Trans., 1B (1988) 663 [see Fig. 8].

[9] J. S. Reed, "Introduction to the Principles of Ceramic Processing," J. Wiley, New York (1988) 246 [see Fig. 15.20]; J. K. Wright, et al., J. Am. Ceram. Soc., 73 (1990) 2653.

[10] J. E. Funk and D. R. Dinger, Ceramic Industry (April 1993) 63. (See also p. 194 of ref. 9.)

With complex clay mixtures, sometimes a coarser or a finer clay of a different inorganic chemical composition is added in order to improve the particle size distribution, going in the direction of a Funk-Dinger or other optimized size range. After doing this, the inorganic percentages of calcium, silicon, etc., might become incorrect for sintering at the standard production temperature used in a factory. If that occurs, a "reformulation" computer program is available that can aid in adjusting the ratio of the clays for a best-case compromise between inorganic chemical requirements and other desired characteristics such as particle size.[11]

5.2 Surface Area Calculations

For the simple case of spherical particles that are all of the same size ("monosize"), the specific surface area can be calculated[12] from the particle size by the geometric relationship

$$\text{Specific Surface Area} = \frac{A}{w} = \frac{A}{V\rho} = \frac{4\pi r^2}{(4/3\pi r^3)\rho} = \frac{3}{r\rho} = \frac{6}{d\rho}$$

where A is the *surface area* in m^2, w is the weight in grams, V is the volume in cc, ρ is the density in gm/cc, r is the radius in μm, and d is the diameter (the *equivalent spherical diameter*) in μm. Fortunately, the dimensions conveniently cancel out so that the commonly used units of density and diameter can be put into the equation directly. As described above, powders with a fairly wide size distribution pack better (because they have both coarse and the fine particles), and they sinter better (because of the good packing and also high surface energy). Even with this distribution of sizes, the specific surface area can still be calculated approximately.

---

[11] R .L. Lehman, Ceram. Eng. Sci. Proc., 16 (1995) 137; P. A. Smith and R. A. Haber, Ceram. Eng. Sci. Proc., 10 (1989) 1.
[12] S. Lowell, "Powder Surface Area and Porosity," Chapman & Hall, N.Y. (1991).

As an illustration of the calculation of surface area even when there is a wide particle size distribution, Fig. 5.4 shows the form in which the automated sedimentation instruments usually report the particle sizes. The differences in heights of the curve at the beginning and end of each cell (such as the heights at 0.1 and at 0.2 μm) were used to obtain a table of data,[13] which were then grouped into bigger cells to give the bottom bar graph in Fig. 5.2. For the central size of each small cell, the previous equation was used to calculate the surface area. As an example, for the 0.1 to 0.2μm cell, the central size is 0.15μm. The density of alumina is about 4 gm/cc. Therefore the specific surface area is $6/(0.15)(4) = 10$ m²/gm.

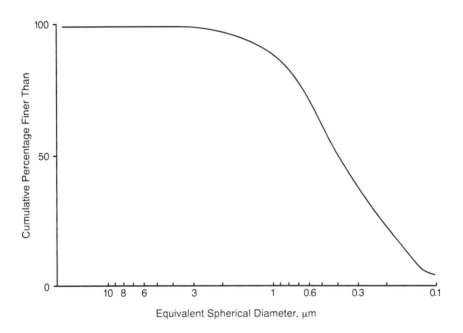

Fig. 5.4 Experimental particle size data for Alcoa A-16SG alumina powder, by sedimentation method.

[13] D. J. Shanefield and R. E. Mistler, Am. Ceram. Soc. Bul., 53 (1974) 416 [see Table III].

The summation of the surface areas for the many different particle sizes taken together is given by the equation

$$\frac{A}{W} = \sum \left(\frac{6}{d\rho} f \right)$$

where f is the weight fraction of powder falling within a certain size range. As an example, for the 0.1 to 0.2µm cell, f is 0.12, so the contribution is $(10 \text{ m}^2/\text{gm}) \times (0.12) = 1.2 \text{ m}^2/\text{gm}$.

The overall summation is approximately what would theoretically be measured by an automated B.E.T. instrument[12] for this powder, assuming spherical particles. This summation has been done in the right hand column of the table,[13] and the specific surface is calculated as being about 6.3 m²/gm. The measured specific surface area by B.E.T., however, was found by the author to be about 12.4 m²/gm, for that particular lot of powder. Many other ceramic powders exhibit similar discrepancies between the surface area calculated from particle sizes and that measured directly, in which the directly measured value is roughly twice as high as the calculated value.

There are several possible explanations for this common discrepancy. First of all, a simple experimental error was unlikely in the cited example, because samples of powder, taken with a spinning riffler,[2,14] had been interchanged with other laboratories, and similar results were obtained. Secondly, care was taken to outgas the sample adequately before doing the B.E.T. measurement, in order to be sure that most of

---

[14] A variety of large and small rifflers (sometimes called "splitters") is available from the Gilson Co., Worthington, OH. Some sort of mechanical riffler is recommended whenever two or more measurements are to be made on the same lot of powder, rather than trusting to the scooping of sublots by hand, which is likely to be unrepresentative. However, care must be taken to use all of the sublot, without further subdividing it by hand. If the riffled sublot is too large, it can not be used. Instead, make several smaller sublots and combine all of each of them, when necessary.

the adsorbed water vapor[15] was successfully "outgassed," and also any dispersant used by the supplier[16] had been removed.[†] Flowing ultra-pure nitrogen was used at $300°C$ for 17 hours. Another possibility was that the alumina particles were rough or porous, but high resolution electron microscopy[17] of the particles indicated that those were not the situations observed here. Still another possible explanation was that the non-spherical powders gave false results in sedimentation, and also in the above calculation. However, corrections were made which took into account the true shapes (distorted tetrahedra), and less than one square meter per gram could be accounted for by this means.[13] Thus the most likely possibility appears to be that there were smaller particles in the finest fraction (0 to 0.1μm) than the sedimentation instrument could measure, and this was contributing as much surface area as all the rest of the coarser powder. In other words, instead of 8% of the powder having a diameter of exactly 0.05μm, possibly 2% of the powder had a 0.005μm diameter (unmeasurable by the sedimentation instrument), which would have contributed 6 m²/gm entirely from the ultrafine region.

---

[15] More water is desorbed at higher "outgassing" temperatures (done just before the B.E.T. measurement), but some still remains on typical powders such as alumina, even at 200°C; for example, see W. H. Wade, et al., J. Phys. Chem., 64 (1960) 1196.

[16] Many ceramic powders are dry ball milled by the supplier, and glycols are often used as "grinding aids" (dispersants) in the mill, leaving residues adsorbed on the surface; for example, see A. Pearson and G. MacZura, U.S. Patent 3,358,937 (1967).

[†] Whenever there are questionable circumstances, the outgassing should be done at a series of higher temperatures and longer times, until a constant value ("plateau," Fig. 4.5) is obtained with the B.E.T. instrument. This will remove most adsorbed layers of dispersant, atmospheric water vapor, and nitrogen, etc. The B.E.T. method works best between 2 and 7 layers of its own adsorbed gas, and previously adsorbed impurities in that range of thickness must be removed beforehand.

[17] D. J. Shanefield and R. E. Mistler, Western Elec. Engr., 15 [2] (April 1971) 26 [see particularly Fig.6]; D. J. Shanefield, et al., in "Ceramic Processing Before Firing," G. Y. Onoda and L. L. Hench, Eds., J. Wiley, New York (1978), 141 [see particularly Fig. 13.2; A-16SG is dense, excellent-quality, single-crystal sapphires].

Later experience with this and other ceramic powders has pointed to the B.E.T. surface area as being the most significant parameter to be measured, when considering the interactions with organic additives. The particle size distribution is a useful characterization parameter also, but it is not as directly correlated to such questions as "How much dispersant should be used per gram of powder?" A close analogy is the determination of how much paint to use on a large spherical water tank. The total surface area is what must be known, not the diameter of the tank. The more area, the more paint, in direct proportion. Similarly, with ceramic powders, the total area is often approximately proportional to the amount of dispersant, lubricant, binder, and other additives that should be used. Even the qualitative questions of "Which families of dispersant and binder should be used?" are answered best by knowing the specific surface area.

An older method for estimating the surface area is somewhat analogous to painting a wall. In this test, a visible dye such as methylene blue[18] or alizarin is adsorbed onto the powder surface by stirring in a dilute solution of the dye. For example, on clay in the presence of alizarin, a white color indicates approximately 5 m^2/gm, a pink color indicates about 9 m^2/gm, and a red color indicates about 12 m^2/gm. This test gives a rough idea of how much dispersant and/or binder to use. When methylene blue dye[19] is utilized in the test, the color scale (which is similar to the above) is called the "methylene blue index," sometimes abbreviated as the "MBI" in the clay industry.

---

[18] For the chemical formula and structural diagram of this and many other organic chemicals, a convenient reference is the catalog of the Aldrich Chemical Company, Milwaukee, WI.

[19] G. W. Phelps, et al., "Rheology and Rheometry of Clay-Water Slips," Cyprus Co., Sandersville, GA (1980), page 240.

6. COLLOID SCIENCE — *AS APPLIED TO CERAMICS*

The study of very fine particles dispersed in fluids falls under the category of colloid science. Definitions of some of the most important terms used in this field appear in Appendix II. It is suggested that the reader look quickly through that word list and get a general idea of the information available there before proceeding further in this chapter.

Ceramic powder particles are mostly larger than true *colloid* particles, and typical slip will settle out somewhat if it is left standing overnight, while a *sol* will not. However, ceramists use the scientific principles of colloid science to disperse their coarse material, just as if the powders were finer. Also, the "fines" in many ceramic powders, that is, the smallest particles, usually are of colloidal size. Whenever they are not effectively dispersed, they can cause trouble in mass production processing, so the principles of colloidal science are important in making sure the fine particles are properly dispersed. Another reason for ceramists to study colloids is that one of the major thrusts of advanced ceramics research is to extend practical processing down into the colloidal range of 0.1μm or so, in order to be able to sinter the newer powders like silicon nitride which are difficult to densify.

6.1 Adsorption

On the scale of atomic sizes, most powder surfaces are somewhat rough, with micro-size pits that might be classified as open pores. Therefore, many examples of *adsorption* onto a surface also involve some *absorption* into these pores. It is probably true that almost all examples of absorption involve some adsorption, as well. However, for defining those two words in Appendix II, each of the chosen examples epitomizes just one phenomenon or else the other, with close to zero overlap of the two concepts. Fine points like these need to be clarified as much as possible, because a detailed understanding of adsorption is very important in ceramics technology, especially with the emerging high performance materials.

Bonding of the Adsorbate to the Powder. Because van der Waals forces can act on all materials, they are often one of the driving forces for physical-bonding adsorption ("physisorption"). Another cause for adsorption is that some powders tend to have electrically charged or at least slightly polarized surfaces, and these will attract any *adsorbate* ions which have the opposite charge. For example, certain clays have an inherent tendency toward a positive charge on the edges of their flat platelet shaped particles, because the broken or strained ionic bonds at these surfaces leave atoms such as calcium behind, and calcium tends to give up electrons and take on a positive charge. On the major flat surfaces of these particles, there is a tendency toward developing a negative charge (see Fig. 6.1), because oxygen atoms are left behind when the particles are broken away from a larger crystal.[1] If the clay particle is placed in a solution of sodium chloride, positive sodium ions would chemically adsorb ("chemisorb") onto the negative flat faces, and chloride onto the edges, forming new ionic bonds.

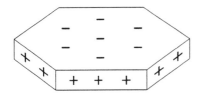

Fig. 6.1 Many types of clay particles are hexagonal platelets with natural surface charges in water, as shown.

Lewis acid/base coordinate bonds can also be the driving force[2] for *chemisorption*. Hydrogen bonds can link the adsorbate liquid or gas to the powder, if both it and the surface contain atoms that can readily be H-bonded together, an example being alcohol adsorbed onto alumina.[3] On silica surfaces, stronger covalent chemical bonds can be formed.[4] Infrared ("IR") spectroscopy is one of the techniques used for proving

[1] H. van Olphen, "Clay Colloid Chemistry, " J. Wiley, New York (1977); G. W. Phelps, et al., "Rheology and Rheometry of Clay-Water Slips," Cyprus Co., Sandersville, GA (1980) 42.
[2] F. M. Fowkes, Advances in Ceramics, 21 (1987) 411 (particularly page 411).
[3] M. D. Sacks, et al., Advances in Ceramics, 21 (1987) 500.
[4] H. Ishida , et al., Polymer Engrng. and Sci., 18 (1978) 128; E. P. Plueddemann, "Silane Coupling Agents," Plenum Press, New York (1991).

whether or not true chemical bonding is involved,[3,4,5] as opposed to van der Waals bonding, which would not be detectable this way. IR can show exactly which pair of atoms is bonded together to cause the adsorption, for example aluminum to nitrogen, versus aluminum to oxygen, etc.

Several special cases are of interest in ceramics processing. In the previous chapter on bond types, it was noted that carbon-carbon double bonds involve π electrons which stick out from the molecule in a manner that allows them to have a Lewis base effect. For this reason, unsaturated organic compounds such as oleic acid become adsorbed strongly to Lewis acid compounds such as those containing aluminum. Therefore oleic acid tends to lie flat on the surface of alumina powder, taking up more of the area than the corresponding saturated compound, stearic acid, would occupy.[6] Along these lines of thinking, oleic acid and oleate esters are often observed to be better dispersants and lubricants in ceramic processing than the corresponding stearates, as will be discussed in more detail in a later chapter. Also, unsaturation in the adsorbate can cause more total amount of it to be adsorbed.[7] Of course, the amount of dispersant which is adsorbed is important in ceramic manufacturing processes, and there is usually an optimum amount, with either too little or too much giving poor results.

There have been many publications that reported the amounts of dispersants or binders adsorbed onto ceramic powders.[8] An important practical consideration is that more dispersant, etc., is adsorbed when there is less solubility in the liquid.[8a,9] Evidently this is because the

---

[5] M. L. Hair, "Infrared Spectroscopy in Surface Chemistry," Marcel Dekker, New York (1967); E. S. Tormey, Advances in Ceram., 9 (1984) 140; R. J. Higgins, Ph.D. Thesis, MIT (1990) [see particularly pages 125, 136].

[6] A. Doroszkowski, et al., Faraday Discuss'n of Chem. Soc., 65 (1978) 252.

[7] H. Koelmans and T. G. Overbeek, Disc. Faraday Soc., 1954 (1954) 52; A. E. Lewis, J. Am. Ceram. Soc., 44 (1961) 233.

[8] (a) D. J. Shanefield, et al, Am. Ceram. Soc. Bul., 53 (1974) 416 [see particularly page 419, and also references 18-24 cited in that article]; (b) E. S. Tormey, Advances in Ceram., 9 (1984) 140; (c) M. D. Sacks, et al., Advances in Ceram., 21 (1987) 495; (d) J. Cesarano, I. A. Aksay, and A. Bleir, J. Am. Ceram. Soc., 71 (1988) 250, 1062; (e) L. Bergstrom, C. H. Schilling, and I. A. Aksay, J Am. Ceram. Soc., 75 (1992) 3305 [adsorbed oleic acid was effectively 2.5 nm thick, and had a 31 square nm area, as reported in Table 1 of this reference].

[9] R. R. Stromberg, et al., J. Res. Nat'l. Bur. Stds., 62 (1959) 71; R. G. Horn, J. Am. Ceram. Soc., 73 (1990) 1117 (particularly page 1126); A. Kerkar, et al., J. Am. Ceram. Soc., 73 (1990) 2879 (particularly page 2880).

adsorbate is less likely to stay in solution if it is less soluble, and so it is therefore more likely to leave the solution and go onto the powder surface. In chemical engineering terms, the fugacity is higher; in physical chemistry terms, the chemical potential is higher. Further examples of this principle will be presented in the discussion on dispersants.

An adsorbate which is slowly being added to a smooth surface in an orderly fashion will eventually build up a complete *monolayer*, and more material will then be added as *multilayers* on top of that. The Langmuir equation[10] describes mathematically the way in which the first monolayer builds up, and the B.E.T. equation[11] describes the additional multilayer buildup from layers 2 through 7. These are useful research tools for dilute solutions, smooth surfaces, and other nearly ideal situations. Practical ceramic formulations, however, are almost always quite concentrated and complex, with rather thick multilayer adsorption[12] and competition for adsorption sites among several components.[13] The author has not found the usual mathematical tools to be workable under these industrial-scale conditions, and the equations do not seem to be followed at high concentrations, but references 12 and 13 provide examples of more optimistic views, in case the reader wishes to pursue further research in these areas.

In predicting the probable effectiveness of a dispersant, lubricant, etc., the dimensions of the adsorbed layer are sometimes of interest. Stearic acid, a commonly used lubricant and dispersant for injection molding and other ceramic shaping methods, has been reported to occupy about $3 \times 10^{-21} m^2$ area when it is adsorbed on a powder surface.[14] That number can be used to estimate how much adsorbate is required to comprise a full monolayer, depending on the surface area available on a given powder sample (and other factors discussed below).

In one particular system, the minimum thickness of an adsorbed monolayer of dispersant was shown[15] by electron microscopy to be 1.5

[10] E. S. Tormey, L. M. Robinson, W. R.Cannon, A. Bleier, and H. K. Bowen, in "Adsorption of Dispersants from Non-Aqueous Solutions," J. Pask and A. Evans, eds., Plenum Press, New York (1981) 121; P. D. Calvert, et al., Am. Ceram. Soc. Bul., 65 (1986) 669.

[11] S. Lowell, "Powder Surface Area and Porosity," Chapman & Hall, N.Y. (1991).

[12] V. L. Richards, J. Am. Ceram. Soc., 72 (1989) 325.

[13] See p. 270 and also K. E. Howard, et al., J. Am. Ceram. Soc., 73 (1990) 2543.

[14] F. M. Fowkes, page 325 in "Treatise on Adhesion and Adhesives," R. L. Patrick, ed. Arnold Publ., London (1967).

[15] M. S. Wrighton, et al., Science, 245 (1989) 845; I. A. Aksay, Ceram. Int'l., 17 (1991) 267.

nm. That number can be used to estimate the maximum possible solids loading that can be achieved when the particles-plus-dispersant are touching each other; note that, even with fairly close packing of the particles-plus-dispersant, the ceramic particles themselves can not touch, and less ceramic material can be in a given volume of green body if the adsorbed layer of dispersant is thick. Although a nanometer or so is rather thin, it can still be a significant limiting factor if the ceramic particles are also small.

The thickness of the adsorbate layer can vary by a large margin, according to the type of adsorbate molecule and also the immediate environment. For example, in an aqueous environment around the particle, polymethylmethacrylic acid used as a dispersant at pH 3.4 coils up into little balls, and the thickness of an adsorbed monolayer of this has been estimated[16] to be 3 nm. The reason for coiling up is that polymethylmethacrylic acid is not very soluble in water, so it is attracted to itself more than to the water, and thus it acts as though it were trying to get away from the water. Probing deeper into the causes of this behavior, the water is actually H-bonding to itself, also, and more strongly than it H-bonds to the partly hydrocarbon polymethylmethacrylic structure. It would require energy to pull the water molecules away from each other, and thus the water partially repels the dispersant. Spheres occupy the least volume possible, so the dispersant coils up into nearly sherical shapes.

When the same material is at pH 8.7 (aqueous, of course), the carboxylic acid groups (−COOH) ionize and stretch out into more nearly straight lines, going radially away from the powder (see Fig. 6.2). The thickness of an adsorbed layer of this, resembling electrified hair, was reported by the same authors to be about 10 nm. The reason for this change at higher pH (in ammonium hydroxide solution, which of course is a Bronsted base) is that the polymethylmethacrylic acid dispersant quickly forms the salt ammonium polymethylmethacrylate. That ionizes readily in water, making negatively charged polymethylmethacrylate ions (and also positively charged ammonium ions, which diffuse at least some small distance away). Since each polymer molecule has many −COO⁻ ionic sites, they repel each other electrostatically, which is the cause of the molecule becoming stretched out almost straight.

---

[16] J. Cesarano, I. A. Aksay, and A. Bleier, J. Am. Ceram. Soc., 71 (1988) 250.

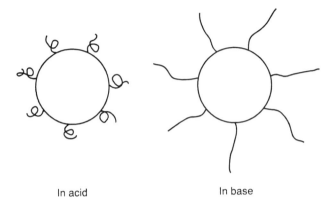

In acid   In base

Fig. 6.2 Adsorbed PMMA configurations. (See also Fig. 3.16.)

Another factor is that these $-COO^-$ ions have a greater tendency to H-bond to water, since they are very strong Lewis bases, and therefore they are not repelled by the water. (Analogously, the salt ammonium polymethylmethacrylate is quite soluble in water, but polymethylmethacrylic acid is only slightly soluble.) Specialists in colloid science sometimes do extensive thermodynamic analyses of the relationships between coiled and uncoiled adsorbates. It is noteworthy that *entropy* plays a part in the calculations also, since the entropic tendency toward randomness is still another factor in causing the molecules to become uncoiled. Thus, at high temperature the uncoiling tendency is usually greater, although there can be exceptions to this if the other factors are also affected by heat.

When larger molecules are used as dispersants in nonaqueous paints, they have been reported to form adsorbed monolayers of approximately 100 nm (0.1μm) thickness,[17] although the measuring techniques were somewhat different from those reported in the previous reference. From these few typical examples it can be seen that the thickness of the adsorbed layers can vary a great deal.

Specificity. The surfaces of crystalline materials, including powders, almost always have "growth steps" and pits on them, as illustrated

---

[17] G. P. Bierwagan, J. Paint Technol., 44 (1972) 46; G. R. Joppien, et al., J. Oil & Colloid Chem. Assoc., 60 (1977) 412.

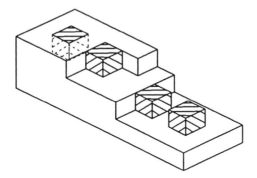

Fig. 6.3 Various degrees of surface contact that adsorbed material can have with a crystal. (Steps can be several atoms high.)

in Fig. 6.3. The sizes of these surface features are sometimes as small as a single atomic layer high, or they can be much larger, even large enough to be visible with the naked eye. They arise from the "bunching" mechanism of crystal growth,[18] or sometimes from irregular cleavage along the crystallographic planes during ball mill grinding. When an atom or molecule from a fluid becomes adsorbed onto the surface of a powder particle, it can be attached at either of the sites shown in the figure. If it is on a relatively wide, flat surface, it will only contact the powder particle on one side. If it is tucked into a step, it will have contact on two sides, and if it is in a "kink site," there will be contact on three sides. If it happens to be exactly the right size to fit into a pit, it could have contact on five sides. The more contact area, the stronger the total adsorption bonding could be. However, only molecules of the correct size can fit into high-contact spaces, while more different types of material could make single-surface contact. It is found in practice that certain molecules, or even parts of a molecule

---

[18] D. J. Shanefield, et. al., J. Electrochem. Soc., 110 (1963) 973; W. D. Kingery, et al., "Introduction to Ceramics," J. Wiley, New York (1976) [see particularly Fig. 10.7].

such as the tail end, will preferentially become adsorbed onto certain sites.

Another factor that leads to the adsorption of only specific molecules is the electric charge at a given site. For example, in the presence of water vapor, silicon nitride reacts slightly with the water to produce some silicate sites at its surface which tend to be acidic, and also some amine or imine sites which are usually basic.[19] A Lewis acid site on the powder surface can attract an adsorbate molecule that is a Lewis base, or vice versa, and a coordinate bond can thus be formed.[20] (A point of critical importance, to be discussed in the next sections, is that these charges can readily change sign, according to the pH of the fluid phase.)

If the molecule being adsorbed onto the powder contains more than one type of functional group, and the powder surface has more than one type of potential adsorption site, then more of the molecule is usually adsorbed. On silicon nitride, such a multifunctional molecule can adsorb onto both the acidic and the basic sites, instead of onto just one of those. This is a fairly general principle, and it is often observed that molecules containing both acidic and basic groups are more readily adsorbed than simpler ones.[21] Two very different monomers or oligomers can be joined together during polymerization to form a *copolymer*, which can contain groups of differing polarity, and these have also been observed to adsorb more readily onto powders in some circumstances.[22]

Another strategy in formulating a very effective dispersant is to physically mix separate molecules (instead of making a single copolymer molecule) of many different types. Some will adsorb onto one type of site on the powder surface, and some on other types, eventually covering most of the total surface and thus being a more effective dispersant than a single compound could be. Since it is sometimes diffi-

---

[19] G. Busca, et al., Mat'ls. Chem. & Phys., 14 (1986) 123 [imine groups]; N. J. Renault, et al., Colloids and Surfaces, 36 (1989) 59 [amine groups, etc.]; L. Bergstrom, et al., J. Am. Ceram. Soc., 72 (1989) 103; S. G. Malghan, et al., Ceram. Trans., 26 (1992) 38.

[20] F. M. Fowkes, Advances in Ceram. 21 (1987) 411 (particularly page 411).

[21] M. Liphard and W. von Rybinski, Progr. Coll. & Polymer Sci., 77 (1988) 158.

[22](a) R. J. Hunter, "Foundations of Colloid Science," Clarendon Press, Oxford (1987) p. 489; (b) R. G. Horn, J. Am. Ceram. Soc., 73 (1990) 1117 [see particularly page 1126].

cult in practice to force even one monolayer of dispersant (or lubricant) to be adsorbed during ceramic processing, these are useful principles to keep in mind for possible improvements in any given formulation.

Hydration of the Surface. Almost all ceramic powders that are in large scale commercial use are either partially or entirely covered with adsorbed water vapor.[23] This has been detected with infrared spectroscopy and other means. Because water is highly polar, and polarity is the major aspect of the van der Waals force, water can stick to most materials. Even relatively inert graphite adsorbs some water in a partial layer,[24] because of an "image force," in which the electron-rich oxygen of the water molecule repels some electrons in the carbon surface, thus "inducing" a slight positive charge, and it is then attracted to the same charge that it induced. This is analogous to a permanent magnet inducing the opposite magnetization in a nearby piece of unmagnetized iron and thereby being attracted to it. Materials more polar than carbon, such as almost all oxide ceramics, strongly attract water by van der Waals forces, usually more than a monolayer. Another cause of water becoming adsorbed onto most oxides is H-bonding. Also, many oxide surfaces will chemically react with water in an ionic manner, making a monomolecular layer of a new hydroxide compound at the surface.

Because of these almost-universal adsorptions of water in ordinary atmospheres, we should consider the surface of nearly all powders to be somewhat *hydrated* surfaces, not just pure oxide or whatever the bulk chemical formula describes. The adsorbed water can both help us to use powders more effectively in ceramics processing and it can hurt us. There are well documented cases where a small amount of chemically bonded water was an anchor for some other material to become strongly adsorbed onto the underlying powder.[25] There are other cases, however, where too much adsorbed water was detrimental

---

[23] D. E. Niesz and R. B. Bennett, p. 61 in G. Y. Onoda and L. L. Hench, "Ceramic Processing Before Firing," J. Wiley, New York (1978) [see particularly p. 73]; S. Masia, et al., J. Mats. Sci., 26 (1991) 2081; R. J. Higgins, Ph.D. Thesis, MIT (1990) 138 [only available directly from MIT, not from the usual UMI microfilm service].

[24] Anon., "ASTM Special Publication 431," Amer. Soc. for Testing Mat'ls., Philadelphia (1967).

[25] H. Ishida , et al., Polymer Engrng. and Sci., 18 (1978) 128 ["silane treatment"]; M. J. Cima, et al., J. Am. Ceram. Soc., 76 (1993) 3136 [sulfonated PVC].

to the ceramic manufacturing process,[26] and these will be discussed further in later sections. The main lesson is that we should be fully aware that our surfaces are at least partly hydrated, usually chemically bonded to the first layer or so, and the next few layers might be physically bonded, if they are present.

6.2 Charged Particles in Suspension

Sources of Charge. If the water reacts to form a new surface hydroxide compound, the formula for the species at the surface can be written MOH, where the M is whatever metal makes up the powder composition. Thus, the M would be calcium in the case of a calcium oxide powder. (The total valence and the other chemical bonds going from the interior of the particle to the surface metal are ignored in this simplified notation.) Some hydroxides, whether bulk or surface, are alkaline (basic), examples being calcia and magnesia. In water, a calcia (calcium oxide) particle will become hydrated at the surface, and then it will ionize slightly, giving up hydroxyl ions which diffuse away, leaving the powder particle surface positively charged. Other hydroxides are acidic, examples being silica and zirconia. Therefore a silica particle in water will become silicic acid, and it will then give up hydrogen ions and take on a slight negative charge.

Titania (titanium oxide) particles in water are nearly neutral. Pure water is, of course, also neutral, but in the presence of a strong Bronsted acid, some of the water molecules will turn into hydronium ions (H_3O^+), taking on positive charges. Hydrated titania, in similar fashion, will become positively charged when placed in an acidic solution. The formula for the species at the surface is sometimes written in the scientific literature as $[MOH_2]^+$ and it is shown diagramatically in Fig. 6.4. This abbreviated notation leaves out some of the bonds to the interior of the particle. In a strongly acidic solution, it should be noted that a given MOH, depending on the tendency to ionize, could either take on a hydrogen ion or lose a hydroxyl ion. The resulting positive surface charge would be the same in either event. The reaction can be found to be written either way in patents and other literature. Similarly, in a basic solution, either hydroxyls from the solution can become adsorbed onto the surface, or the material will lose hydrogen

---

[26] J. A. Casey, Advances in Ceramics, 21 (1987) 439 (slip was too viscous when slightly wet).

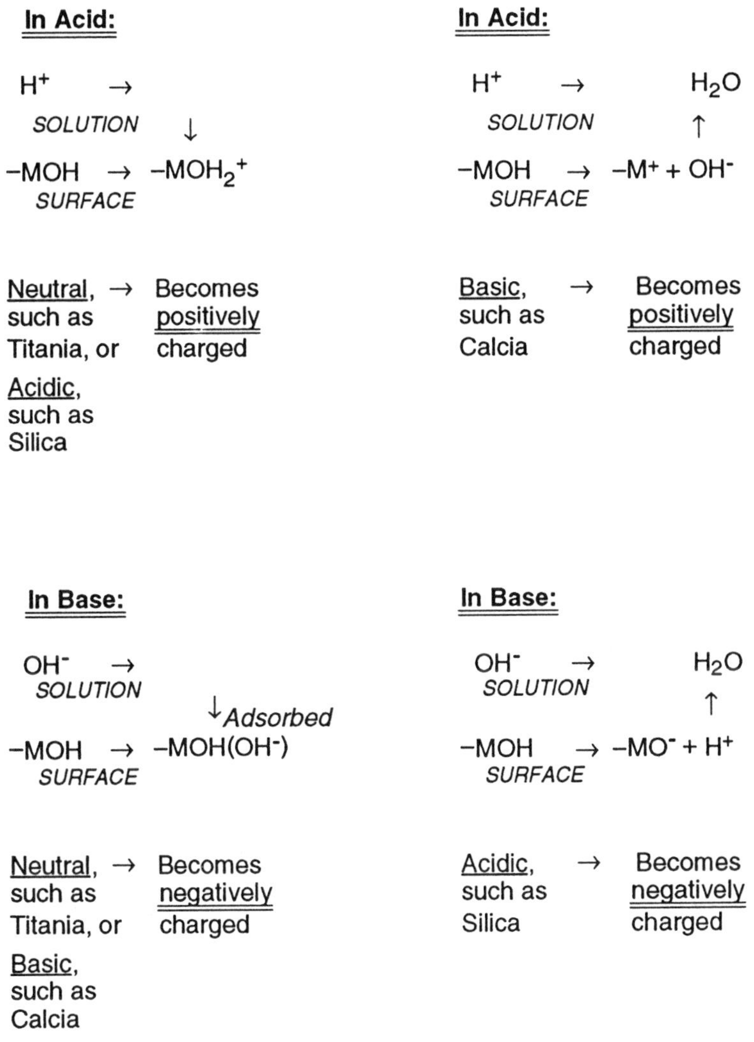

In Acid:

H^+ →

SOLUTION ↓

−MOH → −MOH$_2$$^+$
SURFACE

Neutral, → Becomes
such as   positively
Titania, or charged

Acidic,
such as
Silica

In Acid:

H^+ → H_2O

SOLUTION ↑

−MOH → −M$^+$ + OH$^-$
SURFACE

Basic, → Becomes
such as   positively
Calcia   charged

In Base:

OH$^-$ →

SOLUTION

↓ *Adsorbed*

−MOH → −MOH(OH$^-$)
SURFACE

Neutral, → Becomes
such as   negatively
Titania, or charged

Basic,
such as
Calcia

In Base:

OH$^-$ → H_2O

SOLUTION ↑

−MOH → −MO$^-$ + H$^+$
SURFACE

Acidic, → Becomes
such as   negatively
Silica   charged

Fig. 6.4 Change in charge at powder surface in response to acid or base fluid phase.

ions, depending on the natural acidity of the hydroxide form of the material.

If the adsorbed water was merely van der Waals bonded, and not chemically reacted to form a true hydroxide on the surface, this physi-sorbed water layer could still become charged as above, resulting in an H_3O^+ surface when immersed in an acid, and an OH^- surface in a base. Therefore almost all ceramic powders, even graphite, when placed in an acidic aqueous solution have some tendency to become positively charged, and in bases they are negative. It is a matter of degree, and the magnitude of the charge can be influenced by many factors.

Ionic materials other than water can be adsorbed onto powders and ionize, thus causing the particles to become charged. Dispersants used in water-based slips are usually of this type, and in fact the polyacrylate ions described above are of this type. If the adsorbate is a molecule that has a large number of ionizable groups, then a very large electric charge can be imposed on the powder. However, too long a molecule can have detrimental effects on dispersion, and this will be discussed in the chapter on dispersants.

The Double Layer. If two conductive graphite electrodes are im-mersed in water, and a slight positive charge is put on one of them, negatively charged hydroxyl ions (OH^-) will congregate around the electrode, partially neutralizing the original charge. Careful probing around this sort of electrode has determined that a somewhat diffuse layer of hydronium ions (H_3O^+) will then collect in the general area around the previously formed layer. This structure, called the *double layer,* is shown in Fig. 6.5. The outermost, *diffuse* region, also contains some hydroxyls, which tend to further neutralize the charges. As an example, a calcium oxide particle suspended in acid will be positively charged, but the effectiveness of that charge will become somewhat de-creased by the formation of a double layer.

Two particles having the same charge will repel each other electro-statically, with a repulsion force proportional to the inverse of the dis-tance squared. On the other hand, there will be a van der Waals force of attraction, which is proportional to the inverse of the distance cubed (or possibly higher exponential functions such as distance to the sixth power, depending on whether the London, Hamaker, or other approxi-mations are appropriate to calculate the forces). As the particles ap-proach to nanometer distances, the van der Waals forces take over, and

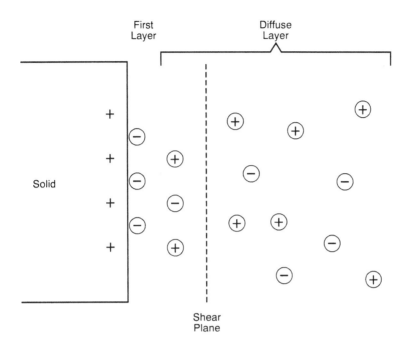

Fig. 6.5 The double layer.

there will be a net attraction until contact is made. This can be seen on the diagram of Fig. 6.6, in which the potential energy decreases at very small distances, similar to a stretched spring relaxing as it contracts. The point of particle to particle contact is called the *primary minimum* in energy.

If the particles are farther away, the van der Waals attraction decreases sharply because of the large exponent of inverse distance, and the electrostatic repulsion takes over, slightly to the right of the hump in the curve. The net effects of attraction and repulsion at various distances were studied by B. Derjaguin, L. Landau, E. Verwey, and G.

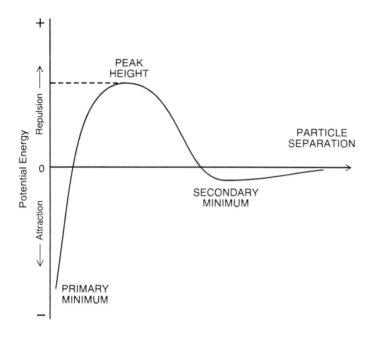

Fig. 6.6 The DLVO curve for two charged particles.

Overbeek, and their combined theoretical calculation is called the DLVO theory.[27] As the particles separate, another point of low potential energy is reached in some cases, and this is called the *secondary minimum*. A pair of particles at this distance will tend to stay there, since energy would have to be put into the system in order for them to become either closer or farther away from each other.

Experimentally, ceramic particles which have been given an electrostatic charge by any of the abovementioned means are often (although not always) found to behave in this manner. That is, there is a stable distance of separation. This distance varies, depending on both the electric charge on the particles and also on the concentration of other ions in the diffuse layer. The order of magnitude of the distance

27 B. V. Derjaguin and J. D. Landau, Acta Physicochim. URSS, 14 (1941) 633; E. J. W. Verwey and J. T. G. Overbeek, "Theory of the Stability of Lyophobic Colloids," Elsevier, Amsterdam (1948).

to the secondary minimum is[28] about 10 nm (0.01 μm). It should be obvious that this distance can vary considerably, depending on the pH and the nature of the adsorbate(s). Simplified DLVO calculations for various conditions also are available, including the Hamaker constants for several ceramic materials.[29] Various other estimates for the effective thickness of the double layer have been reported, usually starting from somewhat different estimates of charge, etc., and resulting in the conclusion that particles should be expected to repel each other at distances of several tens of nanometers separation,[30] although both smaller and larger distances have also been reported. Measurements of the effective distances[31] have confirmed some of the calculations.

One of the largest effects on the dimensions of the double layer is the concentration of charges from other ions which are dissolved in the solution. If there are many such ions, or the ions have multiple* charges (such as Ca^{++} or SO_4^{--}), the diffuse layer will neutralize more of the charge repulsion, and this is said to *collapse the double layer.* The repulsion effect will be diminished, and more pairs of particles will then make contact and form agglomerates. For example, fine clay which is in colloidal suspension in major rivers such as the Mississippi and Nile commonly forms agglomerates when the river runs into the salty ocean and the sodium and chloride ions collapse the double layer. These enormous amounts of clay then settle to the bottom and form the well known deltas. Many mineral nutrients for growing food crops are also present, as well as previously colloidal organic matter ("humus"), which makes valuable, continuously re-fertilized farmland. Possibly the very beginnings of human farming and thus the beginnings of pros-

---

28 D. J.Shaw, "Intro. to Colloid and Surf. Chem.," Butterworth-Heinemann, Newton, Mass. (1992).

29(a) R. G. Horn, J. Am. Ceram. Soc., 73 (1990) 1117; (b) R. J. Pugh and L. Bergstrom, "Surface and Colloid Chem. in Advanced Ceram. Processing," Marcel Dekker, New York (1993).

30 (a) R. E. Johnson, et al., Advances in Ceram., 21 (1987) 323 [10 & 20 nm, Figs. 11,14 and also p. 328]; (b) J. A. Casey, et al., Advances in Ceram., 21 (1987) 439 [10 to 20nm, Figs. 4 &5]; (c) R. Hogg, Advances in Ceram., 21 (1987) 467 [60nm, Fig. 1].

31 W. A. Ducker, Z. Xu, D. R. Clarke, and J. N. Israelachvili, J. Am. Ceram. Soc., 77 (1994) 437.

* The fact that multivalent ions have more effect here is called the Schulze-Hardy Rule.

perity-borne civilization arose at the Tigris-Euphrates delta, all because of this effect.

Another example, closer to ceramics, is the traditional slip casting of clay in plaster of Paris molds, which is enhanced by the same effect. The plaster of Paris is calcium sulfate, and some divalent calcium ions from the mold dissolve in the slip, slightly collapsing the double layer. The slip then agglomerates to some degree, causing the deposited clay body to be more porous than it would otherwise be. Water can easily diffuse through the highly porous first-deposited clay, allowing more clay to also deposit quickly and build up to a practical thickness. In fact, if there is not sufficient calcium ion, the cast occurs too slowly and can soon be self-limiting, not allowing enough deposited thickness to provide a practical level of green strength. When modern porous-polymer molds are used instead of plaster, some divalent ion like calcium is usually added to cause a carefully controlled amount of agglomeration.

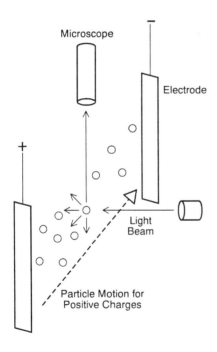

Fig. 6.7 Apparatus for measuring mobility of charged particles suspended in liquid.

Some scientists[32] have recently been experimenting with purposeful additions of ions such as ammonium and chloride in large amounts, in order to not fully collapse the double layer, but broaden it somewhat. This tends to keep the ceramic powder particles slightly farther away from each other, decreasing the tendency toward agglomeration. In fact, if highly acidic solutions are used (pH 2), the colloidal stability is so much enhanced that even after drying the green body can be forced to flow in a plastic manner, without cracking. This is facilitated if ultrasonic vibration is used to agitate the system. The DLVO predictions of potential energy versus distance are not followed in such a system, and it is termed[32] by the authors "non-DLVO."

It is sometimes mentioned in the literature that the double layer thickness is roughly proportional to the inverse of the dielectric constant of the fluid,[29a] and the dielectric constants of water and toluene are 80 and 2.4, respectively. Therefore, in solvents such as toluene, one would expect the particles to be held farther apart than in water, and this has been suggested to explain the excellent dipersion results obtainable in toluene. However, the author has not found this principle to be of great practical value in ceramics, because ionization is weak in most organic liquids. Therefore the peak energy of repulsion in Fig. 6.6 is apparently too low to prevent agglomeration by itself, although it can be of some value when combined with other effects. In other words, some other mechanism instead of electric charge repulsion must be used in nonaqueous solvents to achieve good dispersion, and attempting to achieve strong charge repulsion (by pH control, etc.) in a nonaqueous solvent system by means of the low dielectric constant is not usually very effective by itself.

Another point which indicates that the effect of the dielectric constant is not usually important is the fact that it is outweighed by the much stronger effects of hydrogen bonding. For example, changing to a nonaqueous solvent with minimal H-bonding possibilities can have a large beneficial effect, even if it has a fairly high dielectric constant. The solvent "ethylene chloride" (which is a common name for 1,2-dichloroethane) has a dielectric constant of 11, but it is equally as good as toluene for dispersing ceramic powders, provided that a nonionizing dispersant such as glyceryl trioleate (to be discussed in a later section) is used in both cases.

---

[32] F. F. Lange, et al., J. Am. Ceram. Soc., 77 (1994) 1047 [pH 2, high NH4Cl, plastic green body].

Measurement of the Charge. If a typical ceramic powder particle is placed in an acid and it therefore develops a positive charge on its surface, this charge can be measured approximately with the simple apparatus shown in Fig. 6.7. The particle will move slowly toward the negatively charged electrode. The velocity can be measured by observing the particle with an optical microscope. Large particles will be visible by ordinary means, but particles smaller than the wavelength of blue light (4,000 Ångstroms, or 400 nm, or 0.4 μm) can only be seen in the optical microscope by light scattering. A light beam from a lamp passes through the suspension of particles, and a small amount of the light becomes scattered at nearly random angles. Only the light which is at a 90 degree angle to the original beam passes into the microscope. Against a dark background, points of light which are smaller than can be resolved by the lenses can still be seen. (This is analogous to distant stars being visible against the night sky, in spite of the fact that their great distances make their dimensions be essentially zero. Even with the most powerful telescopes, they could not be seen except as light sources against a dark background.) The points of light are observed against a measuring scale, and this motion is timed in order to measure a velocity. Calibration of the equipment with standardized powders[33] aids in getting consistent results in interlaboratory comparisons, which is otherwise somewhat difficult to accomplish.

Motion of charged colloidal particles due to an electric field is termed *electrophoresis.* The velocity per electric field is the *electrophoretic mobility.* As a charged particle moves in the stationary liquid, the innermost part of the double layer moves with the particle, but the outermost part remains behind. It is apparent that there must be a *shear plane* between them. In the diagram of Fig. 6.5, the potential (or *voltage*) at the shear plane interacts with the voltage between the stationary electrodes, causing the motion. This voltage is defined as the *zeta potential,* and in the older literature this is used as a measure of the effective magnitude of the net charge on a colloidal particle. The zeta potential, ζ, can be calculated from the Smoluchowski equation:

$$\zeta = \frac{v\,\eta}{E\,\varepsilon}$$

[33] J.-F. Wang, R. E. Riman, and D. J. Shanefield, Mat'ls. Res. Soc. Sympos. Proc., 180 (1990) 293.

where v is the velocity, η is the viscosity, E is the electric field (i. e., the voltage per distance between the electrodes), and ε is the electric permittivity (related to the dielectric constant). The zeta potential is a very approximate estimate of the peak height on the DLVO curve.

The use of the zeta potential as a characterization parameter has fallen out of favor with colloid scientists. It is difficult to exactly define where the shear plane is, since its position varies according to the velocity, viscosity, and other factors. Because the potential is different at various parts of Fig. 6.5, movement of the shear plane inward or outward from the particle will change the calculated value of the zeta potential. In practice, it is cumbersome to specify the velocity, viscosity, etc., each time a zeta value is reported. Therefore, another parameter is used more often in modern colloidal work. This is the *electrophoretic mobility*, μ, defined as follows:

$$\mu = \frac{v}{E}$$

Under conditions of constant velocity, viscosity, etc., the zeta potential and the electrophoretic mobility are proportional, and some researchers still prefer to use the zeta potential. Instead of an optical microscope, other methods such as the Doppler effect can also be used to measure the particle velocity, and the entire operation can be computer-automated. Ordinarily, only very dilute suspensions can be studied when light is used as the detecting radiation. The utilization of sound waves instead of light can extend the usable range of solids loadings up to the high levels that ceramists need to study for practical work,[34] and this technique is called *acoustophoresis*.

Commercially, electrophoresis is used directly for manufacturing, as well as for scientific measurements. Cabinets for refrigerators, washing machines, and stoves can be coated with either porcelain enamel or epoxy paint particles by liquid sol electrophoresis, prior to the thermal fusing of the coating. The metal cabinet is simply given a positive charge and immersed in a suspension of negatively charged ceramic porcelain particles or organic epoxy particles, with an inert counter-electrode. Automobile bodies are beginning to be painted by this means, also. An advantage of this method is that, once a thin but complete coating is made, it no longer conducts electricity, and further

---

[34] J. F. Kelso, et al., J. Am. Ceram. Soc., 72 (1989) 625.

thickening stops. For this reason the coatings have exceptionally uniform thickness, with practically no pinholes. In some cases, particularly the coating of file cabinets and office tables, an *aerosol* is used instead of a water or glycol sol, and this method is really a type of electrically controlled spray painting. Practically zero coating material is wasted, and both air pollution and water pollution are minimized.

Changing the Charge. If a fine alumina powder is dispersed in water, and the pH is varied, the electrophoretic mobility (and the zeta potential) will vary, according to a curve similar to the diagram in Fig. 6.8. In the acidic region of the curve, the particles will be take on positive charges and will migrate toward the negative electrode. As the

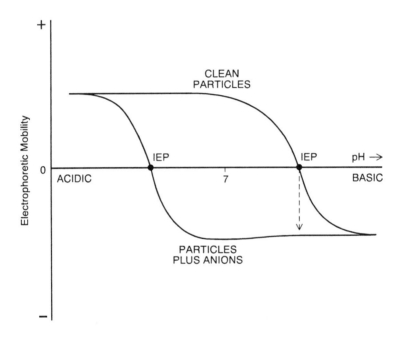

Fig. 6.8 Mobility curves for particles in suspension.

acidity is increased, the velocity will increase, until a saturation value is reached, where the charge on a particle is so great that mutual repulsion discourages any further adsorption of hydrogen ions. At high values of charge, near either end of the curve, the particle will repel other particles, which tends to prevent contact and agglomeration. In other words, the peak in the curve of Fig. 6.6 is both higher and wider than at lower values of pH-induced charge. This is the basis of many practical sols used for ceramics processing; alumina is often made acidic, but strong acid would attack clay, so clays are often made alkaline to stabilize their ceramic slips.

There is some pH value for which the particles have no charge (or, at least, equal positive and negative charges), and this pH is called the *isoelectric point (IEP)*. Another term for the same thing is *zero point of charge (ZPC)*. Different materials have IEPs at different pHs, but the curves have similar shapes. Acidic materials like silica have lower IEPs, and basic materials like calcia have high IEPs. At the IEP, the particles are no longer mutually repelling, and some degree of agglomeration ("coagulation") occurs, with a great increase in viscosity. In some cases the sol becomes a *gel*, or else a soft and somewhat fluffy *floc*, or even a single and rather hard lump.

In addition to hydrogen ions and hydroxyl ions causing the surface of a particle to be electrically charged, other ions can also become adsorbed and change the charge, sometimes in a powerful manner. For example, either of the aqueous dispersants in Table 1.1, sodium silicate (inorganic) or ammonium polyacrylate (organic), when dissolved in water will generate highly polar anions (negatively charged silicate ion or polyacrylate ion). These anions will adsorb onto clay or other ceramic powder, producing a negative charge on the powder particles. Each anion has several electron-rich oxygen atoms sticking out from the structure, and they can polarize an image charge in the surface of the powder particle, causing van der Waals bonding and thus making multiple anchor points for adsorption.

Inorganic molecules such as silicate (and phosphate) become temporary polymers when in solution, although the molecular weights are limited to lower values than the organic materials can have. The acrylates used in industry are usually polymers (polyacrylates). The silicate and polyacrylate dispersants that have ionizable groups like $-COOH$ or $-COO(NH_4)$ on the molecules are called *polyelectrolytes*. When dissolved in water or alcohol, each molecule has more than one ionized group like $-COO^-$ along the length of the chain, and their solutions can conduct electricity; hence the name *electrolyte*. This class of compound finds considerable use in colloidal stabilization technology, because of the multiple charges per molecule. They are also used in very large quantities for paints, water purifying systems, etc., outside of ceramic processing.

A polyacrylate ion of molecular weight 10,000 can have about 100 negative charges.[16] (With a lower molecular weight of polyacrylate ion or similar dispersant, the induced charge on the particle is experimentally found to be less, as expected.[35]) The singly charged sodium and ammonium cations, in contrast, do not have multiple attachment points, and also they are tightly surrounded by water, so these cations are not strongly adsorbed onto the powder and therefore do not give it any significant positive charge.

Because the powder particles have been made more negatively charged, the curve of Fig. 6.8 is shifted downwards, until the saturation line is reached. Most examples in the literature are of this nature,[36] but multiply charged cations such as yttrium ion can also be adsorbed and provide a positive shift in the curve.[37] (In the case of Reference 37, the cation was from a sintering aid, not a dispersant.) Besides van der Waals bonding, Lewis acid/base anchoring of the ion to the powder can also take place,[31] as can any other type of bonding.

Since most powders have positive surface charges in Bronsted-acidic solutions, and many of the multiply-charged dispersant ions have negative charges, simple positive-negative charge attraction causes enhanced adsorption in acid. Therefore more mass of the anion will become adsorbed in acid than in neutral water or in base. This has been demonstrated many times for dispersants,[38] and even for adsorbed binder (higher molecular weight polyacrylate).[39] Therefore, if there are any advantageous effects which could only be caused by such acidity-enhanced adsorption, then these effects should be expected to be greater in somewhat acidic solution. However, many dispersants are strongly attracted to the powder by van der Waals effects, etc., at any pH, so this particular acidity effect is slight in those cases.

A limiting factor at very low pH is that the ionizable adsorbates that are commonly used in ceramics, such as sodium silicate or ammonium polyacrylate, are the salts of weak acids. This is another way of saying that they are not ionized very much at low pH. Thus, if a stronger acid

---

[35] L. Romo, et al., Disc. Faraday Soc., 42 (1966) 232; T. Allen and R. M. Patel, J. Colloid and Interface Sci., 37 (1971) 595.

[36](a) W. R. Cannon, et al., Advances in Ceram., 21 (1987) 525; (b) E. M. DeLiso, J. Kowalski, and W. R. Cannon, Adv. Ceram. Mat'ls., 3 (1988) 407; (c) R. J. Pugh and L. Bergstrom, "Surface and Colloid Chem. in Advanced Ceram. Processing," Marcel Dekker, New York (1993).

[37] J.-F. Wang, R. E. Riman, and D. J. Shanefield, Ceram. Trans., 26 (1992) 240.

[38] A. Foissy, J. Colloid & Interface Sci., 96 (1983) 275; J. Cesarano and I. A. Aksay, J. Am. Ceram. Soc., 71 (1988) 1062; Z. C. Chen, T. A. Ring, and J. Lemaitre, Ceram. Trans., 22 (1991) 257.

[39] K. Nagata, Ceramic Trans., 26 (1992) 205 [see Fig. 3].

like HCl is added in order to force the pH down to about 4, hydrogen ions will go back onto the adsorbate molecules, forming non-ionized species. That effect explains why the charge becomes positive again in strong acid. However, at high pH, much of the adsorbate is ionized .

When a polyelectrolyte dispersant is present, the optimum pH for preventing re-agglomeration of powder particles (in other words, for "stabilizing the suspension") is usually nearly neutral, between pH=6 and pH=8.5. In that range, there is a compromise[40] between the amount of polyelectrolyte adsorbed, as mentioned on the previous page, and the amount that is ionized, as in the previous paragraph. This provides "charge repulsion," but also some"steric hindrance," as below.

6.3 Stabilized Suspensions

Charge Repulsion. Random agitation of the particles due to heat (Brownian motion) results in pairs of particles approaching each other, as well as separating. Particles that have enough kinetic energy to overcome the potential peaks will either make contact and form an agglomerate, or they will separate. A suspension of particles having relatively small electric charges will have low repulsion, and in any given time period a certain fraction of the particles will have enough thermal-motion kinetic energy to approach all the way to the primary minimum. They will then stay together, and agglomerates will grow. As the temperature is raised, a critical point is reached where this process overwhelms repulsion-stabilization for some colloidal systems, and this is called the theta temperature, or θ, which is the temperature of the onset of fast agglomerate growth.

For a suspension to be fairly stable at room temperature, the height of the potential energy peak next to the primary minimum must be at least ~25 millivolts,[41] because the thermal energy available for Brownian motion corresponds to approximately 25 mv at 20°C. The ionic dispersants which are added to ceramic compositions during processing usually can provide a repulsion peak on the DLVO curve of more than 25 mv when they become adsorbed onto the powder. The higher the peak, the more stability can be imparted, with other factors held constant. Higher zeta potentials usually mean more stable suspensions, although this is only a rough correlation. The Brownian motion energies of individual particles vary according to a Gaussian bell

---

40 J. Cesarano and I. A. Aksay, J. Am. Ceram. Soc., 71 (1988) 1062 (Fig.5).
41 R. Hogg, Advances in Ceram., 21 (1987) 467 [Fig. 3].

shaped curve, and therefore a few particles will be out on the "tail of the curve," with higher kinetic energies than the rest. These few can get over the potential hump and form agglomerates even with a peak of 25 mv, and so continuing agglomeration can occur at a slow rate in fairly stable suspensions. However, for powders that are well protected from agglomeration by a high potential peak, agglomeration can be vanishingly slow.

In older literature, the process of forming primary minimum agglomerates is called *coagulation*. The formation of weaker agglomerates where particles reside in each others' secondary minima is sometimes called *flocculation*. However, some of the newer literature does not follow this standard, and there is no clearly defined terminology for these items. In general, a *floc* is a loosely bound agglomerate, usually having a low packing factor.

In addition to knocking pairs of particles toward each other, Brownian motion of the liquid molecules is also what keeps the dispersed particles from settling by gravity to the bottom of the container. The molecules of the fluid phase repeatedly collide with particles of the dispersed phase from all angles, and randomization tends to keep any net motion downward from becoming significant. As agglomerates get larger, they are less affected by collisions with the lighter liquid molecules, and their cumulative settling action becomes greater. In other words, larger agglomerates (or larger individual particles) settle faster. There is no easily calculated cutoff point for the size above which there is significant settling, because the interactions of several bell shaped curves are involved, but approximately 0.1 µm is a rough approximation of this size. Since the median sizes of clay particles range generally from 0.5 µm for fine ball clays to 5 µm for coarse porcelain clays,[42] they will mostly settle out of suspension slowly, even if a dispersant is used. However, the tail ends of their size distributions can extend into the sub-tenth micrometer region, and the finest fractions sometimes do not settle out. This is true for most other ceramic powders, as well. At any rate, even the coarser particles will settle much more slowly if charge repulsion is optimized.

A possible problem which sometimes arises is that particles which have been given negative charges in order to create a fairly stable suspension might have to be mixed with another suspension having positively charged particles. An example is the use of negatively charged

---

[42] G. W. Phelps, et al., "Rheology and Rheometry of Clay-Water Slips," Cyprus Co., Sandersville, GA (1980) 139.

ceramic particles with an adsorbed coating of polyacrylate dispersant, mixed with a positively charged emulsion of lubricant prior to dry pressing. In some cases the negative and positive charges will attract in an uncontrollable manner, causing long chains of flocced material and a great deal of porosity. Usually the solution to this problem is to find two suitable dispersants with the same charge. However, if the concentrations are low enough, sometimes only pairs of particles will stick together and not cause any problems. For example, Cannon, et al.,[36a] by carefully controlling the pH and other parameters, intentionally brought alumina and zirconia particles together in order to make transformation toughened ceramic.

Fast stirring can act like enhanced Brownian motion, thus causing coagulation. This is what happens in a butter churn.

Steric Hindrance In most nonaqueous solvent systems, there is little or no ionization possible, and ionic repulsion is usually a minor factor. Instead, another type of sol stabilization is observed. This involves the *mechanical* prevention of two particles approaching each other closely enough to fall into the primary minima and become an agglomerate. If a uniform coating of a very non-polar organic compound is adsorbed around each of two powder particles, and the particles then move toward each other until the coatings make contact, the particles will not stick but will bounce away from each other. Two reasons for not sticking are: (1) the non-polar materials have much less van der Waals attraction than the highly polar oxide powder surfaces do, and (2) the coatings are weak and will break easily even if they do adhere. This phenomenon is called *steric hindrance.* The word steric comes from the ancient Greek word for solid, and it refers to size and shape being the important factors. This is related to the word "stereo' as in stereophonic sound, meaning simulated three dimensional perception. It is not related to "stearic acid," which comes from the ancient Greek word for fat and is distantly related, through the even more ancient Sanskrit, to the word "steer' for cow or ox.

The organic additives that become adsorbed and provide steric hindrance are analogous to the worn-out rubber tires that are mounted around tugboats to cause the boat to bounce harmlessly off docks and barges. (In England, steric hindrance is sometimes called *elastic stabilization,* which might further encourage one to think of the rubber tire analogy.) Of course, the thickness of the adsorbed coating must be sufficient to prevent van der Waals attraction operating directly from particle to particle, and this is usually the case. However, a small degree of attraction can sometimes operate through the coating.[30a]

In order to be adsorbed onto the powder, these coating molecules often have one polar end for van der Waals or other bonding to the particle, and one non-polar end to prevent sticking to the other particle's coating. If the molecule is too small, it is easily pushed out of the way and is therefore ineffective.[6,43] If it is too big, it can tangle up with the adsorbate on another powder particle and actually cause flocculation.[44]

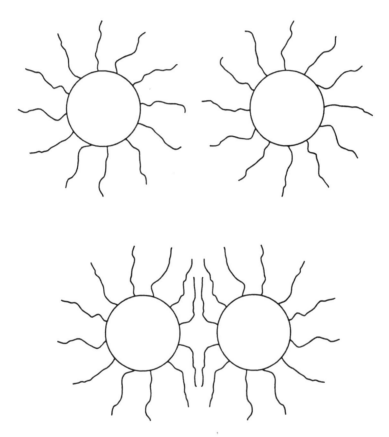

Fig. 6.9 Decrease in entropy as sterically hindered particles approach.

---

[43] D. J. Shanefield, et al., Am. Ceram. Soc. Bul., 53 (1974) 416 (Table 5).
[44] See "flocculant" in indexes of ref. 22a and ref. 28.

These molecules are more effective when they are partially stretched out,[45] and they are waving around somewhat randomly because of thermally induced motion. Then, in order for two powder particles to come into contact, the adsorbate molecules would have to be forced to one side, into a smaller and more restrictive space (Fig. 6.9). This would decrease the entropy of the system, and since entropy tends to increase (unless it is overwhelmed by some other action), a closer approach is thermodynamically unfavorable. For that reason, this particular type of steric hindrance is called *entropic stabilization.* An increase in the slip temperature will cause more energetic waving around, and more resistance to being forced into a smaller space. Therefore entropically stabilized systems are more stable when hot and can flocculate when cooled below the theta temperature.[22a, 28] The adsorbate is more likely to be stretched out if the non-polar end has a great affinity for the solvent. On the other hand, if the solvent does not readily wet the adsorbate molecule, the latter will most likely be coiled into a small ball and will not be effective for entropic stabilization.

From a thermodynamic point of view, another type of steric hindrance to agglomeration is termed *enthalpic stabilization.* This is more likely to be observed in polar solvents. It arises from exothermic bonding of some sort between the adsorbate and the solvent. If two powder particles were to closely approach each other, the strongly wetted adsorbate would have to be squeezed to the point where it sheds some solvent, and this would require heat input, or an enthalpy increase; hence the name. An increase in temperature above the theta temperature causes flocculation. Although these heating and cooling phenomena are of interest theoretically and are described in more depth by References 22a and 28, they are rarely of practical interest in ceramic processing, since slips are not strongly heated or cooled, except in injection molding. Possibly they will be of use in future developments.

A great deal of other theoretical work on steric hindrance has been discussed in the literature of colloids and ceramics,[46] along with experimental support.[47] In real ceramic systems, these model behavior patterns are often mixed. That is, aqueous systems can be mostly dispersed by charge repulsion but still partly by steric hindrance,[22a,28] and

[45] K. Nagata, Ceramic Trans. , 22 (1991) 335 (see Fig. 1).

[46] D. H. Napper, "Polymeric Stabilization of Colloidal Dispersions," Academic Press, London (1983). R. G. Horn, J. Am. Ceram. Soc., 73 (1990) 1117.

[47] S. S. Patel, Annual. Rev. Phys. Chem., 40 (1989) 597.

nonaqueous systems can have mostly steric hindrance but still some charge repulsion, particularly if a trace of water or some alcohol is present.[48]   The mixed mechanisms are sometimes called *electrosteric,* a term used mostly in England, or *semisteric.*[49]   Unfortunately, some of these words such as semisteric and coagulation are not used the same way by all authors.   For example, semisteric can sometimes mean slowly settling and not quite stabilized steric hindrance effects.[50] Practical examples of each of these topics will be covered in the chapter on dispersants.

6.4 Viscosity

The science of *rheology* is the study of viscosity.   Figure 6.10 shows diagramatically some of the various types of *viscometer* instrument used for measuring the viscosity.   The viscometer used most often in the ceramics industry is the cylinder type, but a minor problem with it is that the shear rates are different at the sides compared to the bottom. The capillary flow method is used with high viscosity samples for injection molding and extrusion, but a problem with this type is that often there is "plug flow" or "slippage" of the sample against the outer walls,[51] causing errors in the apparent shear rate.   (This usage of the word "slippage" is not related to the ceramics word "slip" in Appendix I.)   The cup method is not used very much in ceramics technology, although it probably should be, because it is closer to some types of industrial forming equipment (extruders, etc.) than are some of the other methods.   An auxiliary piston can be used with compositions having very high viscosities.   There is no slippage against the walls, but the effective shear rate is difficult to calculate, although it can be obtained experimentally by running the same sample in this and also in another

---

[48] F. J. Micale, et al., Disc. Faraday Soc., Number 42 (1966) 238; R. J. Pugh, et al., Colloids and Surfaces, 7 (1983) 183.

[49] W. R. Cannon, , et al., Advances in Ceram., 26 (1989) 525.

[50] F. F. Lange, et al., J. Am. Ceram. Soc., 77 (1994) 922.

[51] I. Tsao and S. C. Danforth, J. Am. Ceram. Soc., 76 (1993) 2977.

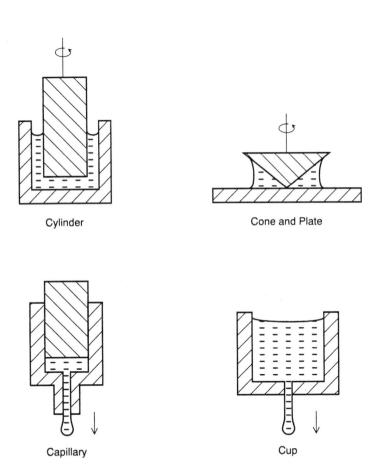

Fig. 6.10 Viscosity measurement configurations.

apparatus. The concepts of shear and slippage are illustrated in Fig. 6.11, and further discussion can be found in the Appendices. The cone and plate method is advantageous because it involves a constant rate of shear across the entire sample: near the center, the circumfrential rate of motion is small, and the distance between the moving cone and the stationary plate is small also. Near the edge, the rate is high, but the distance is proportionately high also.

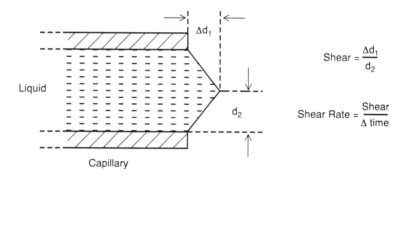

$$\text{Shear} = \frac{\Delta d_1}{d_2}$$

$$\text{Shear Rate} = \frac{\text{Shear}}{\Delta \text{ time}}$$

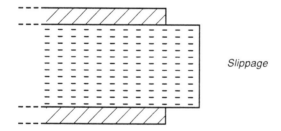

Slippage

Fig. 6.11 Shear concept, and possible slippage error with high viscosity materials for extrusion, etc. Slippage is also called "plug flow" (see section on lubricants in Chapter 9).

Another type of viscometer, not shown in the diagram, involves oscillation[52] instead of continuous rotation. The equipment can be made in various configurations. A computer is required to interpret the results, but it can be an inexpensive microprocessor. The apparatus can be small and easily used in a factory environment. Another advantage of this type of system is that it can detect "viscoelastic" behavior in ceramic slips. This means that there is mostly viscous (that is, frictional, heat-losing, plastic) flow, and also a small amount of elastic (spring-back, non-lossy, rubbery) flow. The mathematical equations describing this are analogous to the mechanical analysis of a brake (lossy) plus a perfect spring (non-lossy), or the electronic analysis of a resistor plus a capacitor. The lossy part is referred to as "real," and the non-lossy part as "imaginary," since the latter involves the square root of minus one. Elastic properties are usually undesirable in ceramic processing, but occasional mysterious results can be explained by the detection of this behavior in the flow characteristics.

When volatile solvents such as MEK are used, care should be taken to ensure that solvent evaporation does not cause the viscosity to increase and/or the temperature to decrease, thus giving misleading results. This can be prevented by enclosing the apparatus. Sometimes the simple expedient of wrapping the slip-holding vessel with flexible plastic film can provide a satisfactory enclosure to prevent vapor escape. When only a small amount of slip is available, it might be too little to saturate the immediate environment with solvent vapor and thus prevent evaporation. In that case the environment might have to be presaturated with vapor by placing paper towels wetted with pure solvent inside the plastic film enclosure for a few minutes before pouring in the slip itself. Also, in some cases the temperature should be measured continuously, especially with highly volatile solvents that tend to lower the temperature, or when using high shear rates that tend to raise the temperature.

The various behavior patterns observed are shown in Fig. 6.12. The *viscosity* is the instantaneous (or, in other words, the incremental) slope on the stress/strain-rate curve. The *apparent viscosity* is the slope of the line from the measured value to the origin. This is what is measured by simple viscometers, but computerized instruments can display the true viscosity also; otherwise the data from a simple instrument can

---

52   C. Gelassi, et. al., "Third Euro-Ceramics Proc., Vol. 1," Faenza Editrice, Madrid (1993) 609.

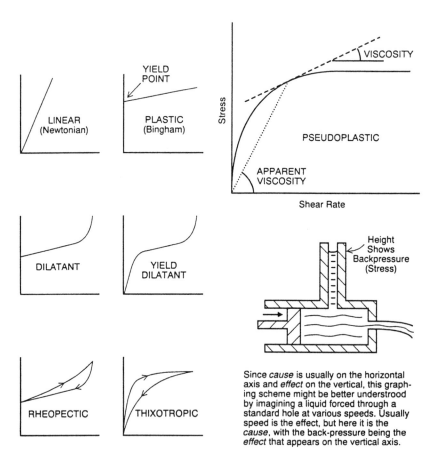

Fig. 6.12 Rheology behavior in slips.

be graphed by hand and the true value obtained geometrically. The *relative viscosity* (not shown in the figure) is the ratio of the viscosity of the whole slip to the viscosity of the pure fluid (usually the solvent plus any other liquids such as plasticizers).

The most desirable pattern for many ceramic processes is *plastic,* in which the material will flow (*shear*) while being molded at high force (*stress*) but not while standing under only 1 g of gravity, awaiting firing. The *pseudoplastic* behavior is an acceptable compromise, and it is often observed in ceramic slips.[t] It is sometimes called "shear thinning," since the viscosity decreases ("thins out") with increasing shear rate. It is often mistakenly called "thixotropic." The pseudoplastic viscosity characteristic is now commonly achieved in paint formulations, so that a thin film can easily be spread along a vertical wall or a ceiling, but the paint will not drip or streak while it is drying. These paints often contain a polyacrylate "latex" (sometimes called an "emulsion" — see Appendix II for further explanation) similar to the binder systems used in ceramics. Many foods also display pseudoplasticity, well known examples being mustard, catsup, and salad dressing. The materials used to obtain the desired flow properties are frequently identical to those used in making water-based ceramic slips, that is, methyl cellulose, and ultrafine silica.

Thixotropic behavior involves *hysteresis,* in which the path on the diagram followed during increasing shear is not followed during the decrease back to zero, and this is quite common. (Broken flocs often take considerable time to re-form.) If the hysteresis is not too extreme, and the "knee in the curve" is sharp enough, it can be quite acceptable. However, it should be noted that the overall behavior is complex and it is quite misleading to characterize the viscosity of most ceramic slips with a single number, although this is often done mistakenly.

The *dilatant* and *rheopectic* patterns are undesirable, because at high shear rates the material becomes too stiff to flow smoothly, and it can crack or even explode a die or mold during extrusion, etc. (Unfortunately, people have been injured by such explosions. The danger in working with a dilatant slip is sometimes underrated.) These last behavior patterns are often called "shear thickening." Still another viscosity behavior pattern is sometimes observed in clay slips, and that is even more complex. It is called "yield dilatancy," and it involves shear thinning in the lower ranges of shear rate and shear thickening at

---

[t] Ceramic slips with desirable yield points and high solids loadings have been discussed by Y. Nurishi, et al., in J. Mat'ls. Sci., 28 (1993) 4456.

very high rates.[53]   This can appear to be pseudoplastic in the viscometer at ordinary shear rates but become dilatant when forced through a small passageway in a pressure casting mold.

Dilatancy occurs at very high solids loadings, where the powder particles are in nearly closest packed arrangements.[54] In Fig. 6.13, horizontal shear of closely packed spheres requires the top layer to ride up over the bottom one, expanding ("dilating") the volume occupied. In experiments with dilatant slips, it is actually found that slight expansion does occur, although it is difficult to observe. Obviously, the viscosity will be very high when particles are forced through this kind of motion, colliding with each other and temporarily changing direction between the horizontal and vertical directions shown in the figure. It should be noted that, as the solids loading is increased until the particles begin to be crowded, it is commonly observed[55] that the viscosity rises in a fairly sudden manner whether the viscosity is "dilatant" or not (to be discussed more in the chapter on dispersants, especially in connection with Fig. 8.2).

Fig. 6.13 Explanation of dilatant behavior.

When dilatancy does occur, one cure for this behavior is to use a lower solids loading. Another type of cure was reported in references 53 and 54, in which the same high solids loading that previously caused dilatancy was kept, but the particle sizes were changed to be more toward the closest packing distribution of Table 5.2. Then more volume of powder could have been fitted into the total slip volume. However, this additional volume of powder was not actually used, and more free

[53] J. E. Funk and D. R. Dinger, Ceram. Industry (April 1993) 63.

[54] G. Y. Onoda, E. G. Liniger, and M. A. Janney, Ceram. Trans., 1B (1988) 611; D. A. Barclay, Ceram. Trans., 26 (1992) 81.

[55] J. W. Goodwin, Am. Ceram. Soc. Bul., 69 (1990) 1694 [see Fig.3].

space (filled with liquid in this case) was available for easier motion, so the dilatancy disappeared. Also, the use of good dispersants and lubricants can often eliminate dilatancy, probably by providing a deformable coating around the particles and allowing easy sliding motions. In many cases a powerful dispersant/lubricant such as stearyl alcohol is the most "robust' solution to the problem; that is, it is least sensitive to lot-to-lot variations in powder, or to random variations in shear rate. If all else fails, the extrusion must simply be done slower (at lower shear rate), to provide sufficient time for the particles to rearrange without riding up over each other and therefore dilating.

Some typical ranges of viscosities used for various ceramic shaping methods have been summarized in Table 6.1. The usable ranges of viscosity quoted in the literature vary a great deal, depending on the shear rate used for the measurement, and the type of shaping equipment used, etc. In practice, it is quite difficult to be sure of the maximum shear rate that is really operative in ceramics shaping equipment. If the flow is very nonlinear, it might be much higher in the center of a tube or orifice than the average shear value calculated for the whole opening would predict. Therefore, the shear rates at which viscosities are measured in industrial quality control laboratories are sometimes unrealistic as estimates of what is going on in the shaping equipment. Great care is needed to estimate an effective shear rate, especially with pseudoplastic slips.

For high concentrations of monosize spherical particles suspended in a relatively inert liquid, the viscosity of the suspension can be predicted approximately by the Einstein equation:

$$\eta_r = 1 + 2.5\,\varphi$$

where η_r is the relative viscosity, and φ is the packing fraction, that is, the volume fraction occupied by the solid spheres. (This equation was suggested by Albert Einstein of relativity fame.) For non-spherical and multisize particles, better fits to experimental data[56] have been obtained with the Mooney-Celik equation

$$\eta_r = e^{\left(\frac{k_1\,\varphi}{1-k_2\varphi}\right)}$$

---

[56] M. Rajala, Am. Ceram. Soc. Bul., 71 (1992) 1817.

Table 6.1. Approximate Viscosities in Ceramic Processing

Process	Viscosity, Centipoise*	Shear Rate, 1/ seconds
Injection Molding and Extrusion	100,000	1,000
Screen Printing of Decorative Ink	40,000	400 to 1,000[‡]
Doctor Blade Tape Casting	7,000[†]	700 to 1,000**
Slip Casting into Porous Molds	700	7

* 1,000 centipoise ("cp") = 1 Pa·sec = 1 N/m^2. Although most ceramists still use centipoise, the Pa·sec is preferred in the "SI" system of standard international units; see Am. Ceram. Soc. Bul., 66 (1987) 1722.

‡ Calculations of shear rates during screen printing have been reported by the following: R. E. Trease, et al., Solid State Tech. (January 1972) 39 [Fig. 2].

† As an example of a typical measurement, the slip described in reference 8a, with a Brookfield HBF viscometer and a #2 spindle, 10 rpm gave 7,000 cp, but 1 rpm gave 30,000 cp. However, other authors report viscosities as much as ten times lower (T. Claassen, et al., J. Eur. Ceram. Soc., 10 [1992] 263) and as much as ten times higher (M. F. Yan, et al., J. Am. Ceram. Soc., 70 [1987] C-280) when tape casting different slips and using different viscometers.

** Some examples of shear rate calculations for tape casting have been reported by W. R. Cannon, et al., in Advances in Ceram., 19 (1986) 161 and in Adv'd. Ceram. Mat'ls., 3 (1988) 374. Since the thickness can be anywhere from 20 μm to 2 mm, the shear rates can also vary greatly.

where k_1 is an empirical constant that varies with different dispersants, and k_2 varies with the particle size distribution. Other modifications of these relationships have been reported to fit certain ceramic slip systems.[57]

The stress at the "yield point" (that is, the knee in the curve) has been estimated in terms of the water content.[58] In another case the stress, τ, at any given point on the curve was related to the strain rate, $\dot{\gamma}$, using the Ostwald power law:

$$\tau = K\dot{\gamma}^n$$

where K and n are experimentally obtained constants, and n turned out to be approximately 0.5. In still another case the slightly more complex Casson equation fit the data fairly well,[59] and it was also noted that the effects of temperature on the behavior followed typical *activation energy* relationships (see Fig. 4.3 and the later section on drying). The basic ideas behind these equations have to do with the crowding of particles that are touching each other at high solids loadings, very much like the explanation of dilatancy. However, in industrial ceramic systems, many other factors can influence the viscosity.

The ratio of two viscosities measured at high and low stresses can be used as an indication of the presence of flocs.[60] Since a floc is a soft and weakly bonded agglomerate, it will break at high stress, leading to a sharply lower viscosity, which can then be taken as evidence for "flocculation." In fact, this is generally thought to be the reason for pseudoplasticity: there are weak bonds (usually H-bonds) throughout the slip, which break at high stress.[tt] Whether such bonds mean that *flocs* or *agglomerates* are present is a matter of rather arbitrary definitions, and there is not any generally agreed upon standard. It should be considered correct to use either word. Some of the weak bonds that can cause pseudoplasticity or even true plasticity can occur

---

[57] R. M. German, J. Am. Ceram. Soc., 77 (1994) 283 [see modified Metzner formula, equation 2].

[58] M. A. Janney and G. Y. Onoda, Jr., Advances in Ceramics, 21 (1987) 615 [see equation 8].

[59] A. I. Isayev, Advances in Ceramics, 21 (1987) 601 [see equation 3].

[60] K. Nagata, Ceramic Trans. , 22 (1991) 335 (see Fig. 3); K. Nagata, Ceramic Trans., 26 (1992) 205 [see Fig. 1].

[tt] It takes time for the broken bonds to "find each other" again and re-flocculate, which causes the hysteresis (thixotropic behavior) mentioned on page 163.

between the powder and the solvent molecules, or between the binder and the solvent molecules, or between binder and other binder. The practical effects on viscosity that can be caused by the additions of solvents, dispersants, binders, etc., will be discussed in later chapters.

An important fact which is made apparent from both theory and experiment is that hydrogen bonding is a major cause of agglomeration and high viscosity in aqueous slips. The use of nonaqueous solvents such as toluene can very much diminish the deleterious effects of H-bonding, because it can not occur in these systems after vigorous ball milling has removed most of the adsorbed water (except to a small degree from residual traces of that water). It is unfortunate that, in spite of the great importance of hydrogen bonding in controlling adsorption, agglomeration, and viscosity, little or no attention is devoted to this phenomenon in most colloids textbooks or in most other courses of study. This also includes the instructional material available in the fields of powder metallurgy, ink, paint, and materials science.

6.5 Wetting

An important situation involving adsorption is the case where two relatively large volumes of material are either weakly or strongly bonded to each other. If at least one of them is a fluid, the bonding (even the weak van der Waals type) can cause it to spread out and cover the surface of the other one. Under some circumstances this is called *wetting*. However, the fluid is also bonded to itself, and there is competition among the various bonds.

Figure 6.14 shows strained bonds at a surface. Because they are already stretched, the bonds are more easily broken than the unstretched bonds inside the bulk of the material. For this reason, surfaces react with foreign substances quite readily. In fact, as was discussed in an earlier section, powder surfaces also react with each other, and this is a major driving force for sintering.

Ionically bonded materials usually have strong internal chemical bonds and also high surface energies. If such material is a liquid, including the case of a melted solid at high temperature, the stretched surface bonds tend to act like an imaginary stretched rubber membrane would act. Since a sphere is the shape having the least volume, the liquid will tend to assume the shape of a sphere, as if a stretched rubber balloon were around it. Gravity, however, tends to flatten the shape against whatever mechanical support is present, so the sphere will be

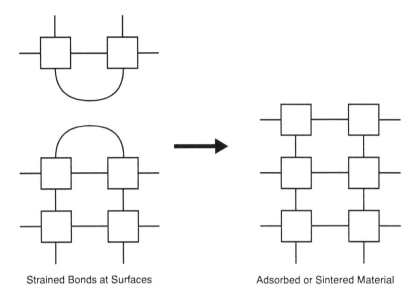

Strained Bonds at Surfaces   Adsorbed or Sintered Material

Fig. 6.14 Strained surface bonds as a driving force for wetting or other adsorption, sintering, etc. Squares represent atoms.

imperfect. Examples would be a drop of liquid mercury (metals have high internal ionic energies) or a drop of water (ionic and H-bonded), sitting on a flat piece of polyethylene plastic. The tendency to act like a stretched membrane at the surface is measurable and is called the *surface tension.*

Organic hydrocarbons such as gasoline have only weak van der Waals bonding between molecules (even though the bonding from carbon to carbon is covalent and stronger), and their surface tensions are usually low. That is, it is easier to stretch out a drop of the liquid material into a long, flat shape. Gravity will flatten a drop of gasoline much more than a drop of mercury.

If a drop of liquid hydrocarbon is put in contact with a higher surface energy material such as mercury, the two will tend to make van der Waals bonds, and possibly stronger hydrogen bonds or other chemical bonds. The hydrocarbon stretches out over (wets) the other surface, but the high energy material will not spread as much because of its high surface tension. This is why oil spreads out to make an monomolecular film on water, but water beads up on oil-covered paper or polyethylene. In other words, oil wets water, but water does not wet oil. The lower surface energy material will spread out on the higher one. (There is also a competition for adsorption by the atmosphere that is present, but this is almost always a minor effect.)

Organic compounds that have polar groups such as −OH on the molecules are intermediate in surface tension. They will wet water but not oil. Higher polarity groups such as −COOH impart an even higher surface tension to the molecule, and organic compounds can be tailored to achieve many varieties of surface properties. A useful option to be discussed later is making one end of a long molecule be polar and the other end nonpolar.

Since water has a fairly high surface tension, it will not wet materials that have even very thin films of oil adsorbed on them. Clay which has been stored in a factory will occasionally adsorb some oil vapor on its surface, invisible to the eye of an inspector. In the process of making a ceramic slip, water will then be found to be repelled by the contaminated clay and will not "wet it out." In that case a *surfactant* must be added to the water, to lower its surface tension and allow it to properly wet out the clay. Some surfactants are anionic (which put a negative electric charge on oil surfaces) and some are cationic (usually quaternary ammonium salts, with large positively charged ions). Further discussion of surfactants will best be postponed until a few other scientific principles are examined.

7. SOLVENTS

In a few ceramic processes such as aqueous clay brickmaking, the water might be called the "solvent," but it does not really need to dissolve anything and only has to fluidize the clay powder during shaping. The fineness of the powder, plus its tendency to hydrogen bond to the water, provides the necessary pseudoplasticity for shaping. Some ultrafine components of the clay are nearly at molecular sizes and can act as the dispersant. (More about that will be discussed in the chapter on dispersants.) Not much strength is needed in brickmaking, since the shape is simple and inherently sturdy.

However, in most modern ceramic processes, it is desirable that the water or other solvent be able to form a true solution of the dispersant and the binder, in order to gain the powerful effects of deagglomeration and strengthening that are available when these materials are broken down to the molecular size level. This and other desirable properties are listed as follows:

(1) Ability to dissolve other additives
(2) Low viscosity at high solids loading
(3) Low tendency to form bubbles during milling
(4) High evaporation rate
(5) Safety, including
    (a) nonflammability and
    (b) nontoxicity
(6) Low cost
(7) Lack of chemical attack on the ceramic powder.

These are to be discussed further in the next sections. Some special attention will be given to the question of how to use water as the solvent instead of various organic liquids. This question is of increasing importance, but there does not appear to be a practical answer for all systems, at present.

7.1 Predicting Solubility

Nonpolar Materials. If two liquids that are similar such as oil and melted wax are placed in contact with each other, they will slowly mix, or in other words, they will dissolve in each other. The driving force is simply the tendency in nature for molecules to move in random directions, that is, the tendency toward an increase in entropy. (For more background on the meaning of entropy, see the dagger footnote on page 102.) In this case the driving force toward mutually dissolving is called "entropy of mixing." The oil and the wax are similar, both being simple hydrocarbons, and both being nonpolar.

At room temperature, the wax is solid, and it will not dissolve in the oil to any large degree. The reason is that the wax molecules are held together by both van der Waals bonding and tangling, and the increase in entropy is not enough driving force to overcome these cohesive tendencies. However, if enough heat is provided to melt the wax, then once the wax molecules are liquefied and therefore free to move, they can dissolve in the oil.

If the solution is later cooled, some of the wax will precipitate out again as a solid. At this lower temperature, not enough molecular motion is provided to keep the large wax molecules moving and to keep them from van der Waals bonding to each other. They are more similar to each other than to the smaller oil molecules, so they line up and are therefore more likely to make multiple van der Waals bonds to each other, and tangle with each other — more so than with the oil.

Materials of Low Polarity. If solid wax is put in contact with warm toluene, but not hot enough to melt the wax, the wax will dissolve even though it has not been melted, which is quite different from the case with oil. The reason why it can dissolve in this case is that the toluene has π electrons from the double bonds which stick out to some degree, and it therefore can polarize an image force in the wax, making stronger van der Waals bonds (discussed in Chapter 2) and thus stabilizing the wax in a dissolved state. In general, solvents containing benzene rings or other double bonds are slightly better solvents for many solids than are saturated solvents. Trichloroethylene is a slightly better solvent than trichloroethane (Fig. 3.2).

Highly Polar Materials. An important principle to remember is that new bonds must be made which are stronger than the old bonds, unless all the bonds are so weak that *entropy* of mixing* is sufficient driving force for dissolving. For example, a solid sugar such as glucose (Fig. 3.23) is H-bonded to itself quite strongly, so it will not dissolve in hot toluene, because its own H-bonds are stronger than any new van der Waals bonds that the solvent can make to it. The entropy of mixing is not sufficient to break the old H-bonds. However, glucose sugar will dissolve easily in water at room temperature, because many new H-bonds can form between it and the water. The moderately strong bonding of the solvent to the solute is called "solvation."

When a new bond forms and atoms thus approach each other, this action gives off energy. That energy could be "dissipated" randomly as heat through a frictional brake. Otherwise, by means of a lever or pulley, it could be stored by using it to stretch a weak spring. The total effect of an energy change (bonds forming, etc.) plus an entropy change (random mixing, etc.) can be quantitatively assessed in terms of a "free energy" change.* This total change must be negative in order for the process to go forward, just like the energy of a stretched spring goes down as it relaxes. The process could be a physical reaction such as melting or a chemical change such as burning. In either case, the change in total *free energy* must be negative when one adds up the thermodynamic values for the starting materials† and compares them for the final products.† The distinction between physical changes and chemical changes is simply the fact that the word "physical" implies weaker bonds. Therefore, the melting or boiling of wax or the dissolving of wax in toluene are all physical changes which involve only very weak van der Waals bonds and tangling. The dissolving of acetic acid into water is really a chemical change, because of the old H-bonds broken and new ones made, and these are chemical bonds (slightly stronger than van der Waals). In fact, the hydrogen is so strongly solvated in water that a new chemical "moiety" forms, the hydronium ion, H_3O^+. (A *moiety* is any new species which was once a part of something else. Usually it refers to something temporary, like an ion or a free radical.) At any rate, the change in free energy is an important criterion for whether or not a process will go forward.

---

* See pages 102 and 103 regarding entropy and free energy. († It is important for the thermodynamics student to understand that the concentration used to calculate "free energy of formation" is usually not unity, because the materials and *products* involved are usually not pure solids. [The product conc'n. is initially ≈ zero!])

It is interesting to consider next the question of why melted wax, whose van der Waals bonds are already mostly broken by heat, will not dissolve in water. The answer, which is quite fundamental to solvency theory but not understood by many engineers, is that in order for the liquid wax to mix in between the liquid water molecules, some of the water's own H-bonds from molecule to molecule must become broken. No new H-bonds from the water to the wax can form and thus provide the energy required to break the "old" ones (this time, in the solvent), so the wax molecules can not enter the water. In other words, the solvent's cohesive energy must be considered also, whenever it is moderately high.

Ionic bonds and coordinate bonds could also be important, if they are broken and remade when a material dissolves, and the same rules apply as described above: new bonds must be stronger than the old ones that need to be broken. Thus NaCl salt is very strongly ionically bonded to itself, but it will still dissolve in water because it then forms just about equally strong new ionic bonds to the water (*solvation*), so entropy of mixing can take over. Ionic bonding requires extremely polar materials, and if they are not highly polar they can not make new ionic bonds. The NaCl is so strongly bonded to itself that it will not dissolve in toluene, because even toluene's π electrons are not polar enough to make truly ionic bonds with sodium or chloride atoms. Following is a summarized listing of a few polarities:

Degree of Polarity	Typical Solid	Typical Liquid
Very low	Wax	Oil
Low	Naphtha-lene	Toluene
Slight	PMMA	Acetone
Medium	PVB	Alcohol
High	PVA, Sugar	Water
Very high	Salt	HCl

Polymers. Ceramists use only a very few polymers that are strongly ionic, and, to the contrary, most of these polymers are only weakly ionic or are slightly polar, like PVA. Obviously, van der

Waals and hydrogen bonds are among the key items to be considered when predicting whether or not one of these materials will dissolve in a given liquid. Polymer chemists sometimes use a set of two or three numbers to predict whether a given solid will probably dissolve in a liquid. These numbers are called "solubility parameters," and they include a scale of polarity, another scale of H-bonding tendency, and sometimes a third scale. The user looks up the two or three numbers for a solute and then finds solvents that have a close match of numbers, and that solute will probably dissolve in that solvent, according to the theory.[1] Various listings are evidently self-consistent when the same experimenter has determined the scales for the solute and the solvent, but they are not always useful to the engineer who just wants to look up PVA, for instance. The m.w., percentage of residual acetate, etc. are rarely specified in the tables, and these might affect the solubility.

Generally, solvents that are chemically similar (for example, both are alcohols or both are ethers) will dissolve in each other. For example, PMMA (Fig. 3.16) contains a *carbonyl* C=O group in its ester linkage, and so does acetone (Fig. 3.4), so that polymer will dissolve in acetone, as long as the m.w. is not too high. Also, the reader should not ignore the fact that the liquid might dissolve into the solid, forming a "solid solution," and this is useful in choosing plasticizers. Thus, PVB (Fig. 3.17) contains an ether oxygen (Fig. 3.8) and so does PEG (Fig. 3.18). Therefore, PEG with a low enough m.w. to be a liquid will dissolve into solid PVB, plasticizing it.

Organic chemicals do not have to contain exactly the same groups in order to dissolve into each other; having similar polarities is sufficient. For example, PVB will dissolve in ethyl alcohol (see table above), because their polarities are similar enough. However, beyond general estimates based on similarities of the main functional groups, plus the overall polarity, more precise solubility relationships are often best left to trial and error experiments.

---

[1] A. Barton, "Handbook of Solubility Parameters," CRC Press, Boca Raton, FL (1983).

If the reader ever has a need for dissolving polar inorganic salts into nonpolar organic solvents, for introducing sintering aids or burnout catalysts into nonaqueous systems, etc., it might be useful to know that almost any cation can be made to dissolve in toluene, etc., by making the trichloroacetate salt.[2] This has been used extensively by chemists for nuclear fuel reclamation, homogeneous catalysis, etc. Also, the basic idea was slightly improved[3] in recent years by the use of fluorophosphates, fluorosulfonates, etc., instead of trichloroacetates, but any of these anions will work in a variety of solvents.

As mentioned in the first chapter, the organic additive systems used in ceramics are conveniently divided into two broad categories, aqueous or nonaqueous. If water is the solvent, then the various organic additives almost always contain OH groups, in order to be dissolvable. (There can be exceptions where the additive is not truly soluble but is emulsified instead. The techniques for doing this will covered in chapters on dispersants and binders.) Making the choice of water or an organic solvent depends on the five criteria listed on the first page of this chapter, and usually all five are considered, with various relative weightings for each one. The second criterion will be discussed next.

7.2 Hydrogen Bonding Effects

Achieving High Solids Loading. Although the molecular weight of water is almost the same as that of methane, water has a higher viscosity than methane which has been liquefied at high pressure or low temperature. The reason is that the H-bonding which is present throughout the water tends to resist flow, while methane can not form H-bonds.[†] Most ceramic powders, being at least partially covered with adsorbed OH groups from the air, will strongly hydrogen bond to water. This tends to raise the viscosity of the slip, for the same reason as in pure water. The more powder, that is, the higher the solids loading, the more the viscosity is raised. Also, the more total powder

---

[2] D.J. Shanefield, J.Inorg. & Nuclear Chem., 24 (1962) 1014.

[3] K. Starke, J.Inorg. & Nuclear Chem., 26 (1964) 1125.

[†] The carbon does not have unbonded pairs of electrons sticking out and available for coordinate bonding to a neighboring hydrogen.

surface area, the higher the viscosity. As illustrated schematically in Fig. 6.14, a point is usually reached where the material becomes a solid and will not flow into the desired shape.

For making ceramics that have low porosity in the finished state, and this category includes most of the newer advanced ceramics, the solids loading in the slip should be as high as possible, so that the shrinkages during drying and firing to full density are minimized. Warpage and cracking will therefore be minimized also, as explained in previous chapters. Because H-bonding is a major cause of high viscosity, the use of nonaqueous solvents such as toluene should allow lower viscosity[4] at a given solids loading; the next step is that the solids loading can then be raised even further until the maximum usable viscosity is reached. With nonaqueous solvents, this maximum occurs at the highest solids loadings now being reported, which have reached about 70 volume % even when using 0.4 μm ceramic powders.[5]

It should be pointed out, however, that aqueous dispersant technology has been improved within the last few years, and the race for the highest solids loading with fine powders is now quite close. Some of the excellent results with water-based systems that have been reported include the successful dispersion of half-micron* powders at well above 60 volume % levels.[6] After drying, and by using pressure casting,[7] the packing can be as high as 68 volume % with the newer water-based systems. Nevertheless, very high solids loadings appear to be easier to achieve reproducibly with nonaqueous systems.

Bubbles and Foam. Because H-bonds are a more energetic form of attraction between water molecules than van der Waals bonds would be, the surface tension of water is higher than that of, say, liquid methane. (The diagrams in Fig. 6.14 illustrate this effect in addition to other surface energy effects.) High surface tension stabilizes bubbles,

---

[4] D. J. Shanefield, Matls. Res. Soc. Proc., 40 (1985) 69; D. J. Shanefield, Advances in Ceram., 19 (1986) 155; P. Nahass, et al., Ceram. Trans., 15 (1989) 355; R. G. Horn, J. Am. Ceram. Soc., 73 (1990) 1117 [page 1128, water effects].

[5] M. J. Edirisinghe, et al., Ceram. Trans. 26 (1992) 165; I. Sushumna and E. Ruckenstein, J. Mat'ls. Res. 7 (1992) 2884.

* "Micron" is an older term for micrometer or μm, one millionth of a meter.

[6] J. Cesarano and I. A. Aksay, J. Am. Ceram. Soc., 71 (1988) 1062; Y. Hirata, et al., J. Ceram. Soc. Japan, 100 (1992) 983.

[7] F. F. Lange and K.T. Miller, Am. Ceram. Soc. Bul., 66 (1987) 1498; B. V. Velamakanni, et al., J. Am. Ceram. Soc., 74 (1991) 166.

if all other factors are equal, so water tends to foam more than, say, toluene when it is ball milled. Even if there is not a great deal of foam in a water-based slip system, any bubbles at all càn be a serious problem in ceramics. For example, if glaze is applied as an aqueous slip before firing, even one stable bubble can ruin the appearance of the fired ceramic. Another example is in thin ceramics for electronic applications, where a single bubble of the same diameter as the thickness of the fired sheet (as small as 20 µm) can result in a hole all the way through. This can become a short circuit if the ceramic is later coated on both sides with metal conductors, which it often is. Additives such as dispersants sometimes have a great effect on foaming, and special anti-foam agents can be added also, as will be covered at the end of Chapter 9. Although bubbles can be removed by various means, it is easier to simply avoid them by using a low surface tension liquid like toluene as the solvent.‡‡

Evaporation Rates for Fast Drying Solvents. Quite obviously, a solvent that is held together with fairly strong H-bonds will require more "heat of vaporization" in order to dry the freshly cast body than will a solvent such as toluene, which is only van der Waals bonded together in the liquid state. Table 7.1 shows some representative values of this parameter, which is sometimes called "latent heat of vaporization" or "enthalpy of vaporization." From the table, alcohol can be seen to have a small amount of H-bonding, and this affects its other properties such as viscosity and surface tension, as well. Since tape casting is an important process for making a huge dollar value of electronic ceramics per year, anything that affects the drying rate during tape casting will have an economic impact. Water as a solvent takes much longer to dry in tape casting, because there must be enough time allowed for the heat energy (that is, calories, not temperature) to get to the slip surface and break the H-bonds.

For some large scale electronic applications such as the tape casting of substrates and IC packages, the dried tape should be about 0.03 inch (0.75 mm) thick. Using toluene as the solvent, a typical tape casting machine has to be about 100 feet (30 meters) long in order for the tape to be dry enough at the end so that it can be continuously peeled off the carrier film and rolled up for storage. With a water-based slip, it would

---

‡‡ Bubble or outgassing (p. 269) problem: H_2O mol=18 cc=22.4L steam (1:1000).

Table 7.1 Thermal Properties of Some Solvents

Solvent	Heat* of Vaporization, Calories/gram	Boiling Point, °C	Flash Point, °C	Explosive Limits, Volume %
Water	540 †	100	——	——
Ethanol	204	78	20	3 – 19
Toluene	95	111	3	1 – 7
Diethyl ether	84	35	– 40	1 – 37
Heptane	76 †	98	– 1	1 – 7
Trichloro-ethylene	57	87	——	——

* This parameter is the enthalpy change, because it is measured at constant pressure. The values listed in most tables are for temperatures at or near the boiling points, but for ceramic drying, they should be measured at the much lower temperature of actual drying. However, any difference is usually insignificant for making decisions about slip solvents.

† For the H-bond, these two numbers may be used to calculate its "bond strength" (as in Fig. 2.1). The H-bonding enthalpy = 540 minus 76 = 0.464 kcal/gm. A mol of water is 18 gm, so a mol has 0.464 (18) ≈ 8.35 kcal/mol from H-bonding. Each water molecule in Fig. 2.2 has *two* net H-bonds, so *one* H-bond ≈ 4 kcal/mol.

not be dry enough, so toluene or methyl ethyl ketone (MEK) nonaqueous solvents are typically used. However, for some other electronic applications such as capacitors and sensors, the dried tape is much thinner, and water is occasionally used in those cases. Of course, water has advantages in safety and low cost, but some ceramic powders are chemically attacked by it.

While the heat of vaporization (ΔH_{vap}) is the main factor in determining the evaporation rate, this is a little bit difficult for engineers to find in reference books, so the boiling point is often used as a rough guideline. However, it is only a very rough guide to the drying rate, as can be seen by toluene's position in the table, in which its boiling point is anomalously high. The heat of vaporization is a better indication, as will be seen from the next discussion.

The drying of wet, freshly shaped bodies almost always goes through two main stages, "constant rate" and "falling rate" (Fig. 7.1). During the first stage, three steps must take place: (1) the solvent diffuses through the body to the surface, and then the solvent waits for

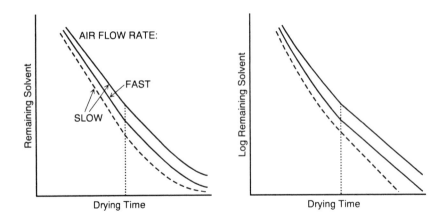

Fig. 7.1 Drying rates for freshly shaped bodies. Dashed line is for higher temperature at the slower air flow rate. In the second stage, after the dotted line, the temperature can be greatly increased without damage (see reference 8, especially Fig. 30.12, and also see article by A. R. Cooper on page 261 of that book).

the right number of calories to diffuse to it via the random motion of warm molecules. (Conductive heat flow follows the same diffusion equations as atoms do.) (2) It then evaporates and (3) is swept away by moving air, or diffuses away. Of the three numbered steps, the second and third are the slowest, because the first step, the diffusion of liquid through a wet material is relatively fast. Whichever is the slowest step in a reaction such as this, chemical or physical (this is the latter), determines the overall speed, just as a slow tollbooth at the end of a short highway tunnel can determine the rate that cars can get through. Getting more heat to the surface, or using a solvent that requires less heat will have large effects on the rate. Also, sweeping the evaporated vapors away faster will have some effect at this stage. In fact, it is useful to nearly saturate the air with solvent vapor, to prevent drying too quickly, which could promote skin formation or cracking. A skin of solid binder sometimes slows down the whole process drastically. Also it keeps bubbles from bursting at the surface, and the skin thus causes large pores ("pinholes") in the fired body.[8a]

In the second stage, the ceramic green body "gels," or becomes stiff, especially if a gelling type of binder is present. The first numbered step above, that is, the diffusion of solvent through the body, is then much slower, and this becomes the controlling step. During this stage, the speed is dependent on both the temperature and the amount of remaining solvent, which each affect diffusion rates. Since the amount of solvent is continually decreasing, the rate of drying also decreases, hence the name falling rate. The slope of the line in the figure is, of course, the drying rate. The mathematics of drying in the second stage is essentially the same as the exponential equation for radioactive decay. Graphing the data on semilog paper makes the second stage plot linear. Equations and data for an example of this have been reported for a tape cast alumina example.[8b] Sometimes there are also other stages in drying,[8c] especially with low-binder, thick, bodies.

Slow Drying Solvents. In the "silk screen printing" of either *decorations* on dishes, or of *conductive* circuit patterns on electronic

---

[8(a)] D. J. Shanefield, in "Encyc. Mat'ls. Sci. and Eng.," M. B. Bever, ed., Pergamon Press, New York (1986) 4855 [particularly page 4856]. [(b)] See chapter by Mistler, Shanefield, and Runk, page 411, in "Ceramic Processing Before Firing," G. Y. Onoda and L. L. Hench, eds., J. Wiley, New York (1978) [particularly Figure 30.11 for an example of *activation energy*]. Briefly, the rate of either a physical or chemical reaction is often: Rate = $Ae^{-(E/RT)}$ where A and R are constants, E is the "activation energy," and T is the absolute temperature (in °K). Plotting log(Rate) vs. 1/T for a process quite often yields a straight line.
[(c)] J. Clausen, et al., Am. Ceram. Soc. Bul., 74 (1995) 40.

substrates, it is desirable that the evaporation be slow. A fast drying solvent such as toluene would dry up while the printing is being repeated, and the screen would quickly become clogged. Here the ink usually includes a glass frit, plus either a *decorative* colored pigment, or an electrical *conductor* for circuit patterns. In either case, for low cost decorations, water is often used because of its moderate drying rate. However, for high quality prints using platinum or gold decorations, and for electronic prints using palladium-silver alloy or gold conductors, even slower drying organic solvents are used.

Two ways to obtain a decrease in drying rate are (1) H-bonding, and (2) higher molecular weight. From the earlier discussion in connection with Fig. 3.1, it can be seen that a progressive increase in m.w. leads to an increase in boiling point, with examples being methane (gas at room temp.), octane (liq.), and octadecane (a solid, mentioned in the discussion but not shown in the figure). Similarly, the discussion in connection with Figs. 3.15 and 3.18 mentioned that oligomers are typically liquids but higher m.w. polymers are typically solids. More tangling and more effective van der Waals bonding in the longer chains encourage the material to remain in the "condensed state" (liquid or solid) and not go to the vapor state. Thus volatile gasoline is made up from alkanes with around seven or eight carbon atoms (see heptane in Table 7.1), nonvolatile motor oil is around C_{10}, wax is around C_{18} or higher, and solid plastics usually have more than 30 carbon atoms in the molecule. Hydrocarbon oils similar to motor oil are the traditional fluidizing liquids in the ink used for printing newspapers, because they do not dry quickly enough to foul the presses. However, hydrocarbon oils are so nonpolar that practically nothing dissolves in them. Therefore more polar liquids with low volatility ("high boiling") are commonly used as ink solvents, and Fig. 7.2 shows the structures of a few common examples.

Another use for slow drying solvents is in ordinary ceramic processes where for some reason flammability is a potential problem, possibly because the factory building is not equipped with explosion proof electric light switches, etc. Chlorocarbon solvents are nonflammable, as discussed in the next section, but suspected toxicity might prevent their use in factories where the ventilation is poor. In those instances, the solvents shown in the figure can be useful. For

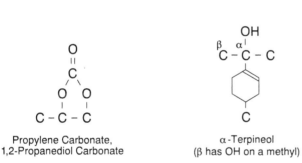

Carbitol,
Diethylene Glycol
Monoethyl Ether,
2(2-Ethoxyethoxy) ethanol

Cellosolve Acetate,
Ethylene Glycol
Monoethyl Ether Acetate,
2-Ethoxyethyl Acetate

Propylene Carbonate,
1,2-Propanediol Carbonate

α-Terpineol
(β has OH on a methyl)

Fig. 7.2 Some slow drying solvents. (Note the similarity between cellosolve acetate, a surprising cause of human stillbirths, versus the commonly used nonionic surfactant in Fig. 8.4, and the plasticizer PEG in Fig. 3.18, all of which are now found in the environment, worldwide. So far, PEG itself has not been implicated, but the nonionic has caused reproduction problems in test animals.)

example, propylene carbonate is a particularly good solvent for many polymers, but it is not well known in the ceramics industry. In fact, if possible chemical reactions with the solvent are a problem, propylene carbonate happens to be unusually inert for a material that even dissolves many ionic salts. It is the main solvent used in very efficient lithium batteries for wrist watches, etc., because it dissolves lithium salts but does not chemically react at the anode or cathode, while almost all other good solvents do. It has been used as a solvent for anhydrous sol-gel ceramic processing without the usual need for water.[9]

The figure also shows cellosolve acetate. The parent compound, which has a hydroxyl group instead of the acetate, was named "cellosolve" because it is a solvent for cellulose, but it is not really a good one compared to methylation of cellulose, and it is no longer used much for that purpose. It and various derivatives do find extensive use as plasticizers, etc., but some serious toxicology problems with these materials will be discussed in the next sections.

7.3 Safety

Flammability. In tape casting machines, static electricity tends to build up to a high voltage as the carrier film is unrolled at the beginning of the process, and the resulting sparks can ignite the vapors of toluene or MEK or other flammable solvents while they are drying. Several very damaging fires have occurred in ceramic tape casting factories where toluene or MEK was the solvent. For this reason, a tape casting process[10] developed by the author and colleagues at AT&T used a nonflammable mixture of trichloroethylene and ethanol as the solvent.* Halogens such as chlorine or bromine, when substi-

---

[9] B.N. Fogelson and D. J. Shanefield, Am. Ceram. Bul., 69 (1990) 541; A. Franco, D. J. Shanefield, and L. C. Klein, Am. Ceram. Bul., 72 (1993) 251 [TEOS was reacted with acetic anhydride in propylene carbonate solvent to make silica].

[10] H. W. Stetson and W. J. Gyurk, U.S. Pat. 3,698,923 (1972) and 3,780,150 ('73); D. J. Shanefield and R. E. Mistler, Am. Ceram. Soc. Bul., 53 (1974) 416, 564.

---

* On a hot summer day, the author was looking forward to the next visit to a patent licensee company, where he was planning to teach them how to use nonflammable TCE instead of their regular toluene. The licensee company, in Colorado, made beer as well as ceramics, and lunch with visitors was in a deep cellar cooled by natural springwater, where visitors could drink all they could hold, and the hosts were known for joviality. Just when the author was daydreaming about cold beer gushing in a cool room, a phone call interrupted to announce cancellation of the meeting — the tape casting plant had burned down. A spark from unrolling polyester film had ignited the toluene vapors.

tuted instead of hydrogen on an organic molecule, tend to inhibit fire. The high temperature chemical reactions of ordinary flames make use of free radicals, but the halogens form competing free radicals which "terminate" the flame reactions by attaching to the ends of the partially burned molecules, preventing further burning. (This is similar to the termination effects of water in some types of polymerization reactions.) Therefore, trichloroethylene does not burn, and mixtures of it and flammable solvents such as ethanol are either nonflammable or very difficult to set on fire without a wick, much like heavy motor oil is.

A commonly used indication of flammability is the "flash point," and Table 7.1 shows data on this for several of the solvents used in ceramic processing. Unfortunately, many factory workers, firemen, and people in the transportation industry seem to misunderstand the meaning of this term. It should be pointed out that the flash point is not the temperature at which an organic solvent will spontaneously catch fire. (The author has had to waste a lot of time convincing people of this negative fact, when new chemicals were being suggested for various processes.) Instead, the term means the temperature at which there is sufficient vapor, generated by evaporation, so that an already existing flame from a match, etc., can cause a fire to start. The test which defines the flash point is done under very narrowly defined circumstances, and the fire does not usually continue after a very short initial "flash" occurs. The test is of some use for very large quantities of solvents, to warn workers not to smoke cigarettes, etc. near leaking containers. For example, ethanol will not catch fire spontaneously if it is held at 20°C, of course, but workers should be careful not to make any sparks near a large container of it at room temperature.

A useful measure of the flammability danger involved with organic solvents is the range of percentages called the "explosive limits," which is also shown in the table. Most organic chemical vapors, for a variety of reasons, will only explode when mixed with air if the ratio of air to vapor is within a narrow range. For example, heptane (see table), one of the major components of gasoline, can not explode in the cylinder of an automobile if 8% or more of the vapor is mixed with 92% or less of air. Instead of running properly, the automobile engine is said to be "flooded" under those conditions. Similarly, if heptane vapor is at a 0.5% level, the mixture is "lean" and also will not explode. On the other hand, if it is 5%, with 95% air, a spark will cause it to explode. Ethyl ether, the material used in cans of the "starting fluid" which is meant to be sprayed into the carburetors of poorly maintained automobiles in very cold weather, has an unusually wide range of flammability, and that is why it is useful for starting engines. Many accidental fires in chemical laboratories and factories have been caused by ether fumes diffusing long distances across floors or tables and being ignited. This is one reason why ether is not used much in ceramic processing, even though it is a good solvent.

Toxicity Regarding Dosage. Let us suppose that the reader is an engineer in a ceramics factory and has recently introduced a new solvent to the production line. Also, a new binder is introduced, and the burnout products have an unusual odor, which the factory workers notice, although they have very little exposure to it because of good venting. But "very little" exposure is not zero, and there is actually some slight exposure to people in the factory, as well as to people living a few city blocks downwind. Suppose, also, that a year later three of the factory workers on that particular production line all develop cancer, but the other 1,000 people in the factory do not. Was this caused by exposure to the new solvent or binder? How can one be sure of the scientific truth involved, and what is the ethical situation in which the reader would be placed if this were a real sequence of events?

Also, can the reader and the employer be sued? Even if it turns out that the suit is without merit, might the employer settle out of court, thus leaving a precedent for the cancer-stricken people to then continue to sue the reader independently, hoping for another million dollar settlement? Since the lawyers would probably get most of the money in a contingency case, a multimillion dollar gain for the plaintiffs would be quite necessary in order for the cancer-stricken workers to get a worthwhile share. These are issues that can not be ignored amongst today's multifarious environmental and legal happenings.

Many ceramics and other chemistry-oriented companies have moved out of various states in the U.S., or even left the whole country, because they could not afford to pay the expenses of meeting recent environmental regulations. This includes the threat of lawsuits, especially in view of the fact that it is easier for workers or neighbors to sue a company if a Government regulation has been violated, whether or not the regulation makes scientific sense. If the companies that have fled this rather complex situation had remained in their old locations, some of them might have gone bankrupt, as many other companies actually have done. Pricing must match the low prices that the competition can charge, working from less developed countries with easier environmental standards. The same situation is true in other industrialized countries such as Germany and Japan, where environmental regulations are newly becoming more strict. Unemployment and decreases in taxable income in the advanced countries are made worse by this loss of manufacturing, among other factors. In those companies remaining in the northeastern states of the U.S., several technical employees have told the author that they spend at least 1/3 of their time on environmental/safety issues, including reading, R&D on new "green" technologies, and filling out government forms.

On the other hand, the governments of the industrialized states and countries must protect their populations from pollution, making use of the most advanced knowledge available. In fact, to some degree they might be protecting the whole world's population. The less developed countries are also beginning to institute severe regulations as their leaders become more knowledgeable, although corruption can often circumvent these regulations because of the generally poor education among voters. At any rate, governments have a grave responsibility to protect, and our past experience tells us that certainly we can not afford to leave our well-being in the hands of profit seeking corporations alone, without any regulation.

These two conflicting factors, the need for protective regulation and the need to employ workers, make it difficult for us to remain unbiased in assessing just how much regulation is ideal, and which regulations are based on scientific truth versus political factors. Indeed, the survival of many companies and even of whole industries depends more and more on a thorough understanding of environmental science. This section and the next one attempt to point out some reports of scientific studies as a background for the reader to use in making environmental decisions on solvents and other organic additives. Some of these reports are evidently not well known to members of the ceramics community who might some day need them as legal precedents, or even to many jurists or juries.

Probably all engineers are familiar with the fact that large enough quantities of any chemical, including water, can be toxic if ingested.

As some biochemists say, "the toxicity is in the dose." The problem of the 1990s is that small quantities of some seemingly innocuous chemicals have recently been found to be toxic, an example being asbestos. In some cases it has taken 30 years of breathing small amounts of asbestos fibers to cause fatal cancer. Now that this is known, it makes the toxicity testing of any newly suggested chemicals extremely difficult. The problem might also apply to other materials, with examples being a new solvent, or possibly a new binder whose burnout products are somehow harmful to the environment, but which might take many years of low level exposure to make its bad effects known.

Until recently, the United States Government often took the position that industry may not expose the population to any chemical that causes cancer, even if extremely small amounts are involved. The famous Delaney Clause in the Food and Drug Administration regulations illustrates that position,[11] whereby nothing may be added to food, *no matter how small the amount,* if it was experimentally found to cause cancer in people *or* animals, no matter how *large* the amount. Although there are separate sets of standards for (a) medicines, (b) food, (c) industrial use, and (d) general pollution, and the Delaney Clause was only applied to food, its extremist philosophy permeated all sorts of other regulations such as the general exposure of school children to asbestos, and also the barely detectable traces of trichloroethylene in well water. This position is now changing among Government researchers, but it still affects many laws that are slow to be changed.

The relevance of this extremist philosophy to ceramics is that a few years ago the U.S. Government claimed aluminum oxide powder to be a carcinogen,[12] and advance announcement has been made that ocean sand will soon be declared to be a carcinogen.[13] All products containing these materials are to be labeled with danger signs, even when they are only meant for industrial use, and even tighter restrictions may be applied later. These announcements were based partly on the fact that crystalline asbestos powder has been found to be a carcinogen when breathed, while glassy powders of the same chemical composition are not. Alumina and sand are both crystalline. Crystallinity is certainly

---

[11] J. Josephson, Environm. Sci. & Tech., 27 (1993) 1466.

[12] Anon., Am. Ceram. Soc. Bul., 68 (1989) 1260.

[13] P. A. Janeway, Ceram. Industry (May 1993) 9.

not the only criterion, however, since many organic solvents such as toluene and 1,1,1-trichloroethane have been announced as likely to be banned for industrial use soon, because their vapors are suspected carcinogens and liver poisons, but only because of ultra-high dosage animal experiments.

An important fact that bears on items such as calling alumina and silica dangerous carcinogens is that all Americans breath asbestos micro-fibers daily, whether the Government regulates this or not. Asbestos fibers were used in automobile brake linings and were worn down into dust for so many years that asbestos dust is now ubiquitous by all the roadsides in the U.S. And yet the unusual "mesothelioma" cancer that heavier doses of asbestos is known to cause is not any more prevalent now than it was before the widespread use of automobiles. Evidently the human body can tolerate very small amounts of asbestos but not somewhat larger amounts. However, the Delaney Clause philosophy that workers must be protected from exposure to *any* amount of carcinogen had dominated the thinking of regulators, and there were few attempts to find a minimum "threshold" amount of exposure that might be allowable within factories. The fact that asbestos by the roadside is a carcinogen and yet it must have a *threshold* has largely been ignored by the regulators.

To provide a perspective on extremely small amounts of asbestos, sand, or toluene, below some allowable threshold, another scientific principle should be considered, and that is the concept of "hormesis." This word is not likely to be found in the scientific dictionaries of most local libraries, or in the vocabulary of most Government regulators, or within the knowledge base of most judges and juries, but it should be more commonly known. The word refers to the fact that some chemicals are harmful or even deadly when ingested in large amounts, but in moderately low amounts they are harmless, and in very low amounts they are even necessary for human life.[14a] An example is iodine, which is illustrated schematically in Fig. 7.3. Just because iodine is a poisonous chemical and must by law be labeled as such when it is sold in drugstores, does not mean that it is always poisonous.

14(a) Anon., Chem. & Engrng. News, (February 6, 1989) 13; [b] G. B. Gori, Chem. & Engrng. News, (September 6, 1982) 25, but see also D. V. Frost, Chem. & Engrng. News, (November 1, 1982) 2; [c] Anon., Chem. & Engrng. News, (February 21, 1977) 25 [regarding selenium].

In fact, if you do not eat a certain minimum amount of iodine every year, you will die. The same is apparently true of niacin, sele-

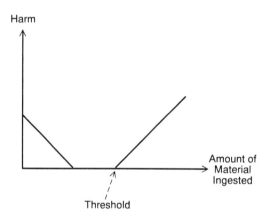

Fig. 7.3 Human body hormesis effects with iodine, selenium, niacin, etc.

nium, and probably many other vitamins and minerals. Of course, the graph in the figure could be drawn for the human intake of water or even air, but iodine becomes poisonous at smaller doses. Selenium has been called[14b] a suspected carcinogen, and niacin is a liver poison in fairly large doses, so all three of these would commonly be called "poisons." But all three of these are "essential nutrients"[14c] and are now contained in most "one-a-day" multivitamin and mineral pills, without any "poison" label. (The selenium is obtained from natural yeast, not a synthetic additive. Aflatoxin is another example of an animal carcinogen[15] which is allowed to occur naturally in food in very small amounts, although it is never added intentionally.) The relevance to ceramics is that large-amount toxicity does not necessarily mean small-amount toxicity, even for suspected carcinogens like selenium.

Toxicity Regarding Animal Tests. Because of the grave importance of decisions about possible toxicity in widespread materials, the Government has relied on testing with animals such as inbred white mice, where the genetic makeup is well established by extensive pedigrees, and the life expectancy of the mouse is short enough so that testing can be done in a reasonable time. In order to obtain test results at lower cost, large doses are given to the animals, for quicker results than trace amounts would provide. (A *potential* bias factor here is that research funding might not be renewed for the next year if the test results just say "everything is OK." Also the news media do not sell much advertising for just reporting "no problems today." Hopefully this is only potential bias and not real.)

Unfortunately, the results of tests done only with animals have been less than satisfactory. For example,[16] the U.S. Government banned both saccharin and cyclamate as a noncaloric substitutes for sugar because of bladder cancers caused in white mice at high doses. Then the tests were repeated and cyclamate was announced to be not a carcinogen after all, but saccharin was still banned. The tests were repeated again by other workers, and the opposite was announced: saccharin was allowed (and still is), but cyclamate was banned. Both were used in other countries, and no evidence of increased bladder cancer or any other problem has been found, even with heavy coffee drinkers who have taken extremely high doses for decades. The U.S. Government has now announced that they are going to revoke the ban on cyclamate in the near future. In other words, every possible combination of results was at one time claimed by the Government: all bad, each one bad and good, and all allowed! The reader does not need to have a Ph.D. in the philosophy of science to recognize unreliable conclusions in this sequence of events.

Alumina powder also had been tested with animals at high doses, but extensive experience with people in mines and factories had not shown any increase in cancer over that of the normal population. Based on

---

15 J. E. Brody, New York Times (March 23, 1993) A1 [at parts per billion levels].
16 J. Brinkley, New York Times (March 2, 1989) A18.

the human exposure data, an organization of alumina manufacturers successfully lobbied in Washington and got alumina "delisted," or removed from the carcinogen list.[12] (It might be useful to the reader to know that this can sometimes be done.)

There are many other examples of animal tests for carcinogenicity being positive but human experience being negative. One of these is the case of dioxin, the chemical made famous by its presence as the supposedly harmful impurity in agent orange defoliant, and having been called[17] by reputable environmental scientists "the most potent chemical carcinogen," based on animal experiments only, where it certainly is among the strongest synthetic carcinogens. Movie stars have raised funds for the benefit of people who were supposed dioxin victims,[18] and large monetary settlements were obtained from class action lawsuits.[19] However, later studies of human exposure, including heavy exposure for long times, showed that dioxin is neither a carcinogen[20] nor a cause of birth defects in people.[21]

As mentioned above, the U.S. Government has announced that toluene and 1,1,1-trichloroethane are "probable carcinogens," and this is based on animal tests. These two solvents will eventually be banned for ordinary industrial use unless stringent precautions are applied to keep workers from breathing any of the vapors. There is no actual schedule for the elimination of these solvents from the workplace, however, and with the steadily increasing caution on the part of Government agencies regarding animal tests, a change from "probable" to "known carcinogen" might never take place. Table 7.2 attempts to summarize some of these arguments. What is happening now is approximated by the right-hand middle paragraph in the table, in which a wider variety of tests is being made, sometimes with humans

---

[17] C. C. Travis and S. T. Hester, Environm. Sci. & Tech., 25 (1991) 815 [see p. 816].

[18] N. Ravo, "Jane Fonda Finds Peace in Her Time," New York Times (August 4, 1988) B1.

[19] A. H. Lubasch, New York Times (April 22, 1987) 14.

[20] K. Schneider, New York Times (May 11, 1994) A1.

[21] F. H. Tschirley, Scientific American, 254 (February 1986) 29; Anon., New York Times (March 18, 1988) A11; N. Wade, N. Y. Times, (July 24, 1994) Sunday Magazine page 10.

Table 7.2 Some Counterarguments Regarding Cancer

Arguments of Environmentalists	Arguments of Industry People
We can not test each new chemical on 1,000 people for 30 years. Therefore we need:	
(1) large doses,	(1) The existance of hormesis with iodine, selenium, etc., shows that there often is threshold of toxicity. Brake asbestos by the roadside is another example of a *threshold*.
(2) with animals.	(2) Too many cases exist by now where animal tests were positive but human experience was negative. Therefore we should use *human* experience from *past* use, as with benzene versus toluene.
There can be no human experience from *past* use with *new* chemicals.	
	For new chemicals, use (1) more animals, with (2) a wider variety of animals (3) and smaller doses, plus (4) comparison of chemical structures. (5) Limited human exposure should then be watched very carefully. Common sense says there is bound to be *some* slight harm from each new thing (like 40,000 deaths/year in exchange for the usefulness of autombiles).
That is too (1) expensive and (2) slow, and you will be the first people to complain about those two things. And who will pay the liability expenses for this "slight" harm?	
	As long as *reasonable care* was taken, with the best knowledge available *at the time,* there should *not be any liability*.

(often from other countries), and neither side is really satisfied with the way things are developing.

Toxicity Regarding Human Exposure. A special situation exists with the solvents benzene and trichoroethylene (TCE), both of which contain carbon-carbon double bonds. The city of Akron, Ohio was the world's manufacturing center for rubber tires, hoses, etc. for decades, and many companies there used benzene, toluene, TCE, and other organic solvents to clean up equipment after use in making rubber products. Few precautions were applied to protect workers from breathing the fumes. Workers at companies in Akron where benzene or TCE were the standard solvents have developed significantly more cancer than workers at plants where toluene, xylene, and some other solvents were used. Benzene tended to cause leukemia, a type of white blood cell cancer. Workers exposed to TCE tended to get cancer or other disorders in the liver.‡

The presence of the double bonds does not seem to be the only factor, since toluene and xylene also contain them. In fact, it is not even a scientific certainty that benzene and TCE are carcinogens, since many other chemical vapors were breathed by the workers in Akron, including smoke from coal-fired boilers, etc. Thus there are two serious problems in attempting to determine toxicity by human exposure:

(1) No single theory can explain why certain chemicals are toxic and other, similar ones are not — for example, methanol is very toxic but ethanol is only very slightly so, in spite of the similarity in structure;

(2) It is unusual that humans who have been poisoned in industrial situations are exposed to only a single potential causative agent, and often there are several possible explanations.

---

‡ The author has been exposed to large amounts of benzene (40 years ago) and TCE (20 years ago), both by breathing the vapors and contacting the skin to the liquids. The dangers were just being discovered at those times. Traces were still detectable in the author's blood for months after the exposures, illustrating the persistence of these "inert" materials. (While no cancer has become evident, many more people would be needed to draw conclusions.)

At one time it was thought that chemically active materials such as vinyl chloride are the carcinogens, and very inert materials such as benzene and TCE should not be. Present evidence indicates that, although this is sometimes true, there are too many exceptions for the basic idea to be relied upon. In fact, some of the more inert materials such as polychlorinated biphenyl (PCB), which were originally chosen for use specifically because of their inertness, are harmful to humans because they accumulate in the body and remain long enough to have some kind of mysterious biological action. Benzene, for example, does remain in the body for long periods of time (see double dagger footnote), while toluene and xylene can be attacked by the body and metabolized because of their methyl groups that stick out from the benzene ring and are not stabilized by resonance. Therefore they do not stay in the body as long as benzene does, and apparently they are not as dangerous.

Regardless of the incompleteness of present day theories, data from industrial or other large scale usage does exist (so called "epidemiological data"), and this often allows a best guess to be made regarding safety. Some cases such as the asbestos connection with an unusual type of lung cancer are clear cut and beyond dispute at this point in time, because moderate doses were associated with high cancer incidence rates, and that particular type of cancer hardly occurs at all elsewhere. Although cigarette manufacturers in the United States dispute a causal association with lung cancer, there is at least one clear cut example where a high rate of an unusual type of lung cancer occurred among people who breathed a great deal of coal smoke, so one type of smoke can cause cancer.[22] Therefore the ceramist certainly should avoid allowing people in or near a factory to breath any kind of binder burnout smoke. Cancer is, of course, not the only potential toxicity problem. For example, women in semiconductor factories who were using cellosolve or related compounds (Fig. 7.2) to dissolve photoresist polymer were found to have many more stillborn children than other women in the factories. This was in spite of the fact that the open containers were "slot vented" (by horizontal slots attached to strong vacuum pumps) and the cellosolve type materials were thought

---

[22] J. C. Chuang, et al., Environ. Sci. and Tech., 26 (1992) 999.

to be safe. Evidently very small amounts of the vapor caused the problem.

Books are available which compile these kinds of industrial experience, plus animal test data,[23] and they can be used to make reasonable decisions. It is the responsibility of the engineer to read about any chemicals that workers are exposed to and make decisions based on current knowledge. Ignoring this knowledge, not disputing ordinary industrial practice when it is wrong, or not disputing governments and extremist environmentalists when they are wrong, could each have various bad consequences. Of all the possibilities, however, the main responsibility lies in the direction of preventing harm to humans or other aspects of the environment, difficult as that may be to determine.

Random Variations in Toxicity Studies. In the hypothetical example posed at the beginning of the section "Toxicity Regarding Dosage" above, it should be recognized by engineers that three workers getting cancer after the introduction of a new chemical does not prove scientifically that the new chemical caused the problem. For one thing, the human body is exposed to many carcinogens that are strictly natural, ranging from yeast and molds and other parasitic growths on food,[24] to various organic chlorine compounds that some plants generate in order to keep competing plants away.[25] Since cancer, mechanical accidents, and other maladies occur with some degree of randomness, considerable thought must go into any attempt to separate true causes from "background noise" random effects. The separation of cause from randomness is a problem in statistical or "stochastic" mathematics and is beyond the scope of this book. However, there is room for a few important points about the interpretation of statistics to be emphasized here.

As the population of the United States has made more and more use of new medical technology, the deaths from pneumonia, early heart attacks, etc. have begun to decrease, and in effect the population now lives long enough to develop cancer more often. Therefore, a greater

---

[23] N. I. Sax, et al., "Dangerous Properties of Industrial Materials," Van Nostrand Reinhold, New York (1993); Anon., "The Sigma-Aldrich Library of Chemical Safety Data," Aldrich Chemical Co., Milwaukee, WI (1990).

[24] B. N. Ames, Science, 221 (1983) 1256; B. N. Ames, ChemTech (October 1989) 590.

[25] G. W. Gribble, Environ. Sci. and Tech., 28 (1994) 310A.

percentage of people are now getting cancer than 50 years ago. However, the age-adjusted ("a-a") cancer rate has not changed significantly, as far as can be determined from medical statistics.[26] Although it is true that some types of cancer have been appearing more often,[27] and this has caused some alarm among environmentalists, other types of cancer rates have been going down, for unknown reasons.[28] It is somewhat surprising that *a-a* cancer rates have not gone up since the widespread use of food additives, insecticides, etc., during the economic and technological boom after the Second World War. In fact, the articles in reference 26 even attempt to estimate the cancer rate (*a-a*) since the beginning of the industrial revolution around 1750, and it does not seem to have changed. While we know that certain widespread synthetic chemicals are carcinogens, so far the overall picture is not as bad as alarmists would have us think.

Popular news media have reported that in New Jersey the total cancer death rate is higher than in other states. However, that is evidently because the total death rate from other causes such as heart attacks and accidents is lower in that state, probably due to helicopter ambulance services, proximity to hospitals, etc. Since everyone eventually dies, a decrease in one cause of death will lead to a statistical increase in another. It turns out that the *a-a* death rate of cancer in New Jersey is indistinguishable from that of other states, in spite of the large concentration of chemical industries there. This example is brought up in order to highlight some commonly repeated misunderstandings about "cause and effect" in this field.*

---

[26] J. C. Bailar, New England J. Med., 314 (1986) 1226; L. G. Lutzker, ChemTech (July 1988) 396; E. W. Volkmann, Chem. & Engrng. News, (February 22, 1988) 2; B. N. Ames, in "Important Advances in Oncology, 1989," V. T. DeVita et al., Eds., Lippincott, Philadelphia (1989).

[27] N. Angier, New York Times (August 24, 1990) A18.

[28] R. N, Hollingworth, Chem. & Engrng. News, (May 8, 1989) 36.

* A parallel example of improved medical treatment involves OSHA safety regulations. The Government claims credit for a lowering of *death* rates in industrial accidents since the OSHA regulations began 20 years ago (B. P. Noble, New York Times, January 23, 1994, page 25). However, the death rates from automobile and home accidents also dropped, because of improved medical treatment. In contrast, the *occurrence* of U.S. industrial accidents has actually increased since OSHA, at the same rate as before OSHA regulations were instituted (D. J. Shanefield, Chem. & Engrng. News, January 10, 1994, page 2). In other words, the statistics do not show definitely that [see next page]

Since cancer and other maladies do occur nonspecifically at random, in addition to being causable specifically by asbestos fibers, etc., it is instructive to examine some of the behavior patterns of random numbers. Let us suppose that among 2,000 workers in a factory, half catch colds at some time during each year, and the other half do not. If ten people who are working in the same room all catch colds, does that mean that there is some specific cause such as low air temperature in the room, or coughing and sneezing transmitting the germs from one person to another? The probabilities can be worked out, but there is a simpler way to consider what might happen. In a table of random numbers,[29] looking at 2,000 numbers, often it can be seen that ten or more odd numbers occur in an unbroken sequence. For example, in reference 29, twelve even numbers occur in row (on line 160), eleven odd numbers in a row (line 193), eight even numbers in a row (line 153), etc. If the reader does not have a random number table but does have a computer handy, the BASIC program

```
10 PRINT INT(2*RND(1))
20 GOTO 10
```

will generate a screen full of zeroes and ones at random. Ten zeroes in a row or ten ones are a fairly common occurrences. The average person, not educated in statistics, would probably guess that ten or twelve heads in a row during a long series of coin tosses means that there definitely is a weight bias in the coin, or that the tosser is cheating. Indeed, a somewhat oversimplified probability estimate of $(0.5)^{10}$ or one chance in 1,024 does seem unlikely to happen. What must be realized is that, although some combination of events is "unlikely" just from random variation, it can still happen from time to time, and not necessarily when expected. In other words, "very unusual random events do occasionally happen." Medical researchers call this

---

[continued from page 197] Government safety rules have *caused* anything beneficial, even though intuitive feelings might indicate that they should have. This is another example of the difficulty in evaluating safety regulations.

[29] W. H. Beyer, CRC Standard Mathematical Tables," CRC Press, Boca Raton, FL (1987) 583.

"clustering," where a grouping of events does not have any known cause but can be explained simply by random occurrences.**

What should be said about probable "cause" if three workers all get cancer while working with a new solvent or while burning out a new binder that has a strongly noticeable smoke odor? The answer depends on the probability of developing that particular type of cancer in the general population, age adjusted. Many judges and juries might, however, be convinced that there must be a real cause in such an event, even if the mathematics does not prove it. While three cancers in the whole factory does not seem to be causative, three at the same tape casting machine and none elsewhere in the factory would probably be grounds for litigation in the minds of many people. The modern engineer should be able to clearly explain the background of such events to judges and juries who are not engineers.

To illustrate with a real example of how judges, juries, the news media, and even the responsible chemical engineers can fail to understand statistics, consider the case of Love Canal. Married women living on top of a buried chemical waste dump reported the occurrence of several children born with cleft palates, and also several stillbirths. The federal and local Governments declared that the chemical waste had been the cause, and all the people in the large housing development were forced to move to other locations. A class action lawsuit was settled with the company that had originally owned the landfill site, even though they had warned all the people involved and the land had been re-sold several times to various intervening owners, including a public school. A large amount of money had to be paid to the people who moved, although it was not nearly enough to cover the losses of their houses. Instead of pointing out that cleft palate and stillbirths occur throughout the world at high enough rates so that this could be a random occurrence, the responsible engineers were strangely silent, perhaps not understanding the whole situation. Years later, several independent studies showed that purely random clustering was the

---

** Should engineers, managers, military leaders, and politicians who have three successes in a row be trusted *without review* on their future fourth decisions? Probably not, since even ten in a row could have been luck.

most likely explanation.[30]   All the government agencies involved then allowed people to move back into the same houses, and no statistically significant problems have occurred so far.   However, general news media and extremist environmentalists still use this case as a prime example of industrial irresponsibility. Most educated people, if asked now, would probably only remember the condemnation of the houses and not the final resolution of the case. Actually, many of us would probably hesitate to move our families to Love Canal now, regardless of what the statisticians say about clustering. The purpose of studying this case is only to understand how randomness can be misleading, and how legal actions are not necessarily fair if engineers fail to explain this.

More recently, a case of landmark importance was hardly noticed by the mass media, because a decision of the "no problems today" category was reached, but it is a clear example of how the courts are slowly becoming more sophisticated about statistics.[31]   Six former workers at a Government nuclear facility developed cancer. They sued the Government for damages, claiming that the nuclear materials *can* cause cancer and therefore *did*. However, the court's decision was that the incidence of cancer at that facility was, although not zero, no higher than in the general population, and the types of cancer also were not unusual. The court expressed sympathy for the former workers but did not grant any monetary settlement beyond ordinary health insurance claims. This case will probably be cited as an important precedent in future lawsuits.

Use First, Test Later. While new chemicals that are meant to be added to food or medicines must be tested before they can be legally used in the United States, this does not apply to new chemicals meant to be used in industrial processes within closed containers. It is assumed that the workers are not going to eat, drink, or breathe the chemicals. Therefore they can be used first, and the "testing" can consist of seeing whether or not any harm is later found among the workers. The use of these new materials in closed containers is not at all regulated, with a few possible exceptions such as radioactive

---

[30] P. M. Boffey, New York Times (May 18, 1983) A1; E. Nieves, New York Times (July 12, 1994) B4.

[31] Anon., New York Times (July 23, 1994) 26 [small article at bottom of page].

materials, explosives, and addictive drugs, which are each covered by special laws.

For comparison, there are two other categories of industrial chemicals that are regulated, as follows.

> Chemicals that are new and which are dumped from the factory into the air that we all breathe or into river water, and which fit into certain broad categories which might cause some environmental harm, such as "volatile organic compounds."

> Chemicals that are not new and in fact are well known to be toxic. An example is benzene, which is not forbidden completely, but which must be kept at lower than certain concentrations in the air breathed by workers.

There is, in effect, an indirect legal restriction on the use of new industrial chemicals: if they are found later to have escaped from containers and to have harmed workers or neighbors, the company and also the responsible executives and engineers can be sued for damages. But this would have to take place after the harm was already done, and testing does not have to be done beforehand.

Now suppose a new industrial chemical is used and later reported to be harming workers somewhere, in spite of reasonable efforts to protect them from exposure. A company can legally switch to a similar material, as long as it is not exactly the same, and this is allowed to be used with absolutely no prior testing. Naturally, the same indirect threat of a future lawsuit would still apply, but there is a great difference between this and the situation for prior testing of food additives and medicines.

This might be of importance to ceramists, even those who use water as their solvent. As briefly mentioned above, female workers in California were using cellosolve acetate (Fig. 7.2) to remove photoresist polymer from semiconductors. (Cellosolve or its acetate ester is occasionally used as a plasticizer for ceramic binders, also.) The vapors were thought to be safe, and also the venting was fairly good, so very little was breathed. After several years of use, the women workers who got pregnant experienced far more stillbirths than in the general population or elsewhere in the same factory.[32] They sued and obtained a very large settlement from the company.

Cellosolve itself is ethylene glycol with an ethyl group instead of one of the end hydrogens. Ethylene glycol alone has a bad combination of two properties: it tastes sweet to people and animals, and it also is somewhat poisonous to both. It is widely used as an antifreeze in automobiles, and accidental spills have resulted in many deaths, particularly of children and pet cats. Propylene glycol has not been thoroughly tested with humans, but it is not poisonous to cats. On the supposition that it is also not poisonous to humans, automotive suppliers have switched over to this as their present and future antifreeze, advertising it as being "relatively safe." Since methyl alcohol and ethyl alcohol are so different in their human toxicities, it is a reasonable guess that ethylene glycol and propylene glycol derivatives also might have different toxicities. However, until a large number of children accidentally drink it, this will not be known quantitatively.

The interesting connection to industrial solvents and plasticizers is that semiconductor manufacturers have switched over to compounds like propoxypropylene glycol acetate, hoping that this is not toxic from the standpoint of birth defects. Suppliers are advertising these variations as "relatively safe." For use as plasticizers or solvents, many companies are now switching to a variety of esters such as propyl acetate[33] and dimethyl glutarate which are found in fruit and other natural foods. However the harmful effects of the cellosolve derivatives were so unexpected that the actual safety of these related materials when breathed, etc., can not be determined except by large scale trial. Therefore, another industrial example of "use first, test later" is now in progress.

Potential Damage to the Broader Environment. In addition to the possibility of poisoning workers in a factory or its neighbors, engineers in the 1990s must consider the possibilities that solvent vapor, or smoke from binder burnout, or gases from burning fuel, etc., might actually affect a whole city, or even the whole planet. Once again, the responsible engineer can become perplexed by the claims of

---

[32] B. Hileman, Chem. & Engrng. News (October 19, 1993) 8; L. Ember, Chem. & Engrng. News (June 14, 1993) 8.
[33] E. M. Kirschner, Chem. & Engrng. News (June 20, 1994) 13.

extremists on both sides of a continuing debate, with many uncertainties in the available data.

In Chapter 3 the discussion on CFC compounds, in connection with Fig. 3.9, mentioned the possible effects of man-made chlorocarbons such as TCE and CFCs on the high altitude stratospheric ozone layer. There is also a possible effect of man-made organic compounds on the lower atmosphere, especially in large cities such as Los Angeles and Mexico City, where rings of mountains surround the cities, preventing heavy gases from being blown away. "Volatile organic compounds" (VOCs) such as gasoline, toluene, etc., which accumulate to some degree in the atmosphere, have been said to interact with sunlight to form ozone. While ozone evidently has some good effects up in the stratosphere, it is definitely bad down where we live and breathe, and its accumulation in Los Angeles has probably caused many deaths from emphysema and asthma, as well as a great deal of extreme discomfort. Various Government regulations have succeeded in decreasing the amount of VOCs in the atmosphere. However, the ozone in Los Angeles and some other cities has not decreased nearly as much as was predicted. There is some evidence that nitrogen oxides (NO_x) are the main culprit, not VOCs, and the emissions of these are beginning to be limited by regulations also. This subject will be dealt with further in the section on binders and their burnout problems.

At any rate, the emissions of VOCs are going to be limited more and more, whether this makes sense or not. Organic solvents will have to be condensed or otherwise prevented from escaping into the atmosphere. Magazines listed in Appendix III carry up-to-date information on the details of such regulations, in addition to advertisements for equipment for condensing and recycling, adsorption on charcoal, catalytic oxidation, and auxiliary flame burning of the vapors. Condensing often requires the removal of co-condensed water before the solvent can be recycled. Sometimes an auxiliary flame is the easiest to implement, by installing a small gas burner in the flue, to make sure that all vapors are burned to carbon dioxide and water vapor.

While the burning of organic compounds to form carbon dioxide might seem to be completely innocuous, environmental extremists have raised the specter of "global warming" due to "greenhouse gases" such as CO_2, especially from the burning of fuel. Because ceramic sintering is such an energy intensive process, any Government regulation on the emission of CO_2 is likely to have a big economic impact on this industry. The carbon dioxide concentration in the atmosphere is quite low, only about 1/30 of that of argon.[34] However, this is increasing, and it absorbs infrared radiation from the sun, which presumably could cause the atmospheric temperature to rise with grave consequences for humanity. Whether the increase is due to mankind or to natural cycles is debatable. Long before the industrial revolution, the atmospheric carbon dioxide concentration was five times higher than it is now, and yet it came down by itself, because of natural cycles.[35]

The atmospheric temperature has also cycled much higher and lower than now, without intervention from mankind. While some scientists report on the basis of polar ice analysis that the temperature is rising dangerously,[36] others say that the ice data is not reliable,[37] the atmospheric water and clouds are far more important,[38] and the temperature is not really rising at all.[39] However, many environmentalists and mass media writers assume that man-made greenhouse global warming is definitely a real danger of grave proportions, and they seem to be ignoring counterarguments. Many countries are now voluntarily trying to limit their carbon dioxide production in order to minimize this, although the United States Government is not,[†] at least for the time being.

---

[34] D. R. Lide, Handbook of Chem. & Phys., CRC Press, Boston (1993).

[35] T. Cerling, Environ. Sci. and Tech., 26 (1992) 1474; D. J. Shanefield, Environ. Sci. and Tech., 26 (1992) 1857.

[36] B. Hileman, Chem. & Engrng. News (April 27, 1992) 7 [see the parallel CO2curves on p. 16]; J. Jouzel, et al., Nature (London), 329 (1987) 403; J. M. Barola, et al., Nature (London), 329 (1987) 408.

[37] C. S. Powell, Scientific American (March 1994) 22.

[38] R. S. Lindzen, Science (1990) 249; R. S. Lindzen, Environ. Sci. and Tech., 24 (1991) 424; K. D, Breyer, Chem. & Engrng. News, (June 3, 1991) 37.

[39] Anon., "Tree Rings Show No Warming," N. Y. Times (July 13, 1993) C5 [during the last 3000 yrs]; K. Leutwyler, Sci. Amer. (February 1994) 24 [last 4 yrs.).

[†] The author sent about 20 photocopies of reference 35 to key people in the Government and atmospheric research. Whether or not this had any actual influence on the country's hesitation to make new regulations on greenhouse gas emissions is indeterminate, but it is a possible example of a citizen's duty in a democracy to aid the flow of any specialized information that is available.

A more likely threat to the well-being of humanity is to be found in another combination of two facts, and again it is not certain whether one causes the other, or whether the cause is really some third unknown factor. First, it is known from laboratory experiments that several chlorocarbon compounds such as polychlorinated biphenyl (PCB) and DDT and CFCs can mimic birds' and animals' sex hormones and interfere with the reproduction of normal offspring. This is also true for materials that are not chlorocarbons, an example[40] being a commonly used nonionic detergent (used for ceramics, see Chapter 8, ref. 12). Since the boom after the Second World War, these compounds have become analytically detectable all over the earth, and they are now present in every human being and all birds that have been analyzed.

The second fact is that populations of almost all frog species and of many bird species are dropping suddenly, all over the world.[41] Of even more import, the human male sperm count has been dropping steadily about 10% per decade for the past 40 years.[42(a)] If the rate continues to be linear, the end[42(b)] of human reproduction will be in 2053 A.D.!

Environmentalists are quick to blame industrial chlorocarbons (TCE, etc.) and certain detergents, and also the chlorination of drinking water, and the stakes are so high that it is difficult to debate the matter calmly and rationally. There appear to be so many different kinds of evidence against chlorocarbons in particular, that an ethical scientist is tempted to feel that one or another of the environmentalists' arguments is bound to be right, so why not give up and join them? Extremist environmentalists are trying to lobby for a ban on all synthetic chlorine compounds, including the chlorination of drinking water.[43]

An analogy that might help clarify the ethical philosophies is the fictional story of a splinter group religion that claims nonbelievers will suffer in hell for eternity if they do not convert. At first sight, it might seem to be a good idea to join this religion, since the stakes are so high:

---

[40] J. Raloff, Science News, 145 (January 8, 1994) 27.

[41] H. Youth, Worldwatch (January 1994) 10; Anon., ChemTech (June 1994) 52.

[42(a)] N. E. Skakkebaek, Brit. Med. J., 305 (1992) 609; B. Hileman, Chem. & Engrng. News, (September 21, 1992) 5; J. Rennie, Scientific American (September 1993) 34; J. Raloff, Science News, 145 (January 22, 1994) 56; [(b)] The science fiction novel by P. D. James, "The Children of Men," Random House (1994), predicted the end to be in 2021 A.D.

[43] J. Holusha, New York Times (December 20, 1992) E2.

punishment for an infinite time. But many other splinter group religions then make themselves known, each claiming that a very different set of behaviors is uniquely necessary to avoid eternity in hell. It soon is apparent that there is an infinite number of possible behavior patterns being recommended, and no way to determine which is correct. There would be no rational way to choose which religion to join.

Returning to the environment, although the stakes are high (the possible destruction of all human life), we don't know which behavior pattern to follow, eliminate the chlorination of drinking water, eliminate nonionic detergents, etc. There is not an infinite number of things being proposed, but there are very many, and some of these eliminations would have high cost. For example, DDT and PCBs were banned in the U.S. at about the same time, mostly because eagles and similar bird species were dying out and there seemed to be a connection. As a result of the DDT ban, malaria has made a serious resurgence in many of the poor countries of the world, where the U.S. had formerly donated DDT supplies. Uncountable thousands of people have died or been incapacitated as a result, and a great deal of farmland has had to be abandoned by nearly-starving people. While this was hardly noticed in America, it has had enormous cost elsewhere, particularly in Africa. Also, new mutation strains of mosquito-borne malaria are now appearing and spreading northward which are not responding to the usual treatments, and these might eventually come to the U.S. In evaluating the total ethical situation, the question might fairly be asked, "How many birds are worth how many people? And which birds versus which people?" To further complicate matters, it has been claimed[44] by a toxicologist that recent laboratory experiments with birds proved that PCBs were responsible for the reproductive problem (eggshell thinning) but that DDT was not a cause and should not have been banned. This has not been widely confirmed and accepted by other researchers, but if it is true, then grievous harm of great magnitude has been done to underdeveloped countries through a well-meaning scientific error. The elimination of chlorine water treatment might become another one, of even greater magnitude and closer to home for Americans. The stakes are high, but so is the level of uncertainty.

---

[44] F. Coulston, ChemTech (November 1989) 660.

Many species of wild animals have been observed to nearly die out but later completely recover, for reasons that are unknown to scientists. Disease, cosmic rays, or unseen competitions with other animals could all be responsible. None of the presently observed die-outs corresponds in time with the introduction of any new chemicals. Human sperm cell counts began to decline before the main buildup of synthetic chlorocarbons in the atmosphere, and it has been amazingly linear, while the buildup has been exponential. Natural sources of chlorocarbons, including many that are toxic, dwarf the synthetics by a factor of 100 or more, mainly being produced by algae and molds in the ocean.[25]

What should the ceramist do about using new solvents, plasticizers, detergents, etc.? The guidelines at the lower right quarter of Table 7.2 are suggested, integrating a variety of information sources.**

7.4 COST

Although water is the cheapest solvent, the advantages of faster drying rate and higher solids loading, etc., that are obtainable with organic solvents sometimes outweigh this factor. A mixture of trichloroethylene (TCE) and ethanol (EtOH) was commonly used[8] as the organic solvent for ceramic processing for more than a decade, because of the good solvency of TCE and its nonflammability. When it became identified as a probable carcinogen, many users changed to 1,1,1-trichloroethane plus ethanol. Then the U.S. Government announced a progressive ban on chlorocarbons (including CFCs, but others also), to be culminated in 1995, so many ceramists switched to toluene in advance of the ban. (Actually, toluene had also been used for many years prior to TCE.) The U.S. Government has announced that toluene is also a suspected carcinogen based on animal tests, in spite of the human experience in Akron being generally favorable. More recently, one researcher has claimed that he has detected

---

** A scientist who recently proved that a silicone product is harmless was so heavily attacked by lawyers trying to collect money for people who used the silicone, that this scientist then advised young researchers to avoid such work! This does not augur well for finding honest answers about future environmental issues. (See G. Kolata, New York Times, May 16, 1995, page D6.)

Table 7.3 Alternative Alkane Solvents*

Gasoline

Hexanes†

Ligroin

Mineral Spirits

Naptha

Petroleum Ether

* The word "solvent" sometimes is taken to mean "organic solvent," as contrasted with water. Of course, water is also a solvent, but this fact seems to be ignored when people say "solvent" and mean only the organic liquids. In paint technology, this material is usually called the "vehicle," although that can include dissolved dispersants, etc.

† The plural, as contrasted with the singular "hexane," indicates that there are various isomers and often various slightly different m.w. species present in the liquid. This is much less expensive than "hexane," because the latter has been laboriously purified.

clusters of human leukemia among toluene users,[45] and that toluene should be banned as an antiknock additive in gasoline, since the fumes are breathed to some degree by filling station workers. Other researchers have not yet agreed that the statistical level of confidence is significant for these data, when compared to background noise, and further study is being carried out, with great interest among gasoline manufacturers.

---

[45] C. Maltoni, Environ. Sci. & Tech., 28 (1994) 306A.

At present, the author would recommend a mixture‡ of alkanes similar to octane, in addition to isopropanol. (Octane is not quite as good a solvent as TCE or toluene, because of the lack of π electrons, so pure isopropanol might sometimes be tried instead of the mixture,‡ since isopropanol is more polar than octane.) If one looks up the price of "octane," it will most likely be prohibitively high. That is because heptane, nonane, and other similar compounds were present in the original petroleum from which it was obtained, and they have to be distilled out of the material that is sold as "octane." Because they are all so similar, this is a slow and expensive process. But there is no need to use pure octane. Instead, any of the much cheaper solvents in Table 7.3 can be used, depending on their boiling points, purity, etc.

Some ceramists use cyclohexane, on account of its fairly low cost. It is made synthetically, not by distillation from petroleum, so it does not have to be laboriously distilled in order to be pure. However, it is more expensive than the solvents in the table, and no better at dissolving solutes, because it does not have π electrons as benzene or toluene would have. The ring structure alone does not make it "aromatic."

Isopropanol is recommended instead of ethanol because it can not be used as a recreational drink, at least not twice. (The first use will not result in death, like methanol would, but such a powerful headache and upset stomach will occur that there will be no second usage.) Therefore it does not require the tax stamp that the Government imposes on ethanol sales. Alternatively, "denatured alcohol" which is intentionally poisoned ethanol could be used, but the poisons are different with each manufacturer and are likely to vary from lot to lot, and they might have some uncontrolled side effects on the ceramic process.

7.5 Chemical Attack On The Powder

Aluminum nitride, ceramic superconductors, some types of relaxor ferroelectric ceramics, and other powders might be chemically attacked by water and their properties degraded. In some cases the powder surface can be protected by adsorbing a monolayer of hydrophobic

---

‡ The reason for the use of more than one solvent will be discussed in the chapter on dispersants.

material onto the surface. For aluminum nitride, both silane treatments (discussed in other chapters) and stearic acid have been used to protect the surface.[46]   However, these thin coatings are easily destroyed by ball milling or other aggressive deagglomeration steps during normal ceramic processing, so only mild mixing treatments can be used.

A very effective method for protecting aluminum nitride when it is used in water, even with ball milling, is to buffer the pH to about 6, which prevents chemical attack.[47] Possibly this idea could be adapted to protect other powders, also. A great deal of other research effort is also being directed at present toward protecting the surfaces of these sensitive powders, so that water can be used instead of the more expensive and possibly less safe organic solvents.

---

[46] M. Egashira, et al., J. Mat'l. Sci. Letters, 10 (1991) 994; T. J. Mroz, Am. Ceram. Bul., 71 (1992) 782; H. Gorter, et al., "Third Euro-Ceramics Proc., Vol. 1," Faenza Editrice, Madrid (1993) 617; E. A. Groat and T. J. Mroz, Ceram. Industry (March 1993) 34.
[47] K. E. Howard, U.S. Patent 5,234,712 (1993).

8. DISPERSANTS

8.1 Tests For Effectiveness

Sedimentation Height. If a ceramic powder with a particle size ranging from about 1 μm to 5 μm is well dispersed (deflocculated or deagglomerated) in a fluidizing liquid (the solvent) and then left to settle by the forces of gravity, after a few hours the particles will form a fairly dense compact at the bottom of the vessel, with a packing factor of about 50% of the theoretical density of the fully sintered ceramic. The solvent lubricates the particles as they fall, letting them pack better than they would in the dry "shaken and settled" example listed in Table 5.1, which only would be about 33% of theoretical density (T.D.). This wet-settled packing factor is approximately as good as in the dry "lubricated and pressed" example in that table, even though only 1 ġ of gravity is the force, compared to 10,000 pounds per square inch or so in dry pressing. The reason for the good packing is a highly effective sort of lubrication that the solvent provides, which allows the falling particles to rearrange themselves, filling whatever pores are close to their individual sizes. The settled condition is shown in Fig. 8.1, compared to the same weight of poorly dispersed (flocculated) powder that only wet-settles to a larger final volume with a lower packing factor. If the powder is very much flocculated, and there is a lot of it in the slip, there might not be any visible settling at all, since the floc can fill the entire container. Various intermediate levels of dispersion, occurring either because the dispersant or the mechanical action (ball milling, etc.) are only moderately effective, will result in various intermediate heights of the powder column after a few hours. Therefore the "sedimentation height" of the settled powder has been used as a quantitative measure of dispersion effectiveness.[1]

A problem can occur when this evaluation method is used for the finer powders that are frequently of interest in making high

[1] T. A. Ring, et al., J. Am. Ceram. Soc., 72 (1989) 1918; I. A. Aksay, et al., J. Mat'ls. Res., 9 (1994) 451.

performance, difficult to sinter ceramics, in which a significant percentage of the material is finer than a few tenths of a micrometer in diameter. For instance, if the same experiments are repeated with a powder having about 10 weight % of the particles between 0.9 μm and 0.01 μm, the finest particles will not settle at all if they are well dispersed. In fact, the better they have been dispersed, the less they will settle, and the *higher* the volume, for fine powders. This test has even been used as a measure in the opposite direction as that described above, where a large value of height means good dispersion.[2] Sometimes the results are difficult to interpret, and the method becomes as much of an art as a science, where the experimenter has to have a feeling for the type of powder involved and what might happen with a variety of solids loadings or of other factors.

Minimum Viscosity. Another measure of dispersion effectiveness is the viscosity of the slip, where the lower the viscosity, the better the dispersion achieved. This is quite often used in scientific studies of dispersants. As the experimenter adds more and more dispersant, if fine powders are being used the viscosity is often found to be at a minimum value when approximately 1 gram of dispersant is added to 100 gm of powder (1 "part per hundred"), because that roughly corresponds to a monolayer of dispersant adsorbed on the powder surface (Fig. 4.4). Of course, the exact amount for minimum viscosity depends on many variables, such as the density of the powder, its specific surface area, the size of the dispersant molecule, etc. Some typical values for half micron powders (roughly $10 m^2/gm$) have been 0.8% dispersant,[3] and 1.2% dispersant,[4] and 1.7% dispersant[5] which provided the minimum viscosities. These reports fit into a remarkably narrow range, even though some systems were aqueous and some nonaqueous, and other aspects of the slips were also very different. When the specific surface area is higher, more dispersant is usually needed, often in a nearly linear relationship.

---

[2] R. O. James, Advances in Ceramics, 21 (1987) 349 [see particularly Fig. 3].

[3] A. Karas, T. Kumagai, and W. R. Cannon, Adv'd. Ceram. Mat'ls., 3 (1988) 374; W. R. Cannon, et al., Advances in Ceram., 26 (1989) 525 [see Fig. 2a].

[4] H. Takabe, et al., J. Ceram. Soc. Japan, 100 (1992) 750 [see Fig. 2]; J. Faisson and R. A. Haber, Ceram. Eng. Sci. Proc., 12 (1991) 106.

[5] R. E. Mistler, D. J. Shanefield, and R. B. Runk, page 411 in G. Y. Onoda and L. L. Hench, "Ceramic Processing Before Firing," J. Wiley, New York (1978) [see Table 30.2 and compare to bottom of Table 30.3].

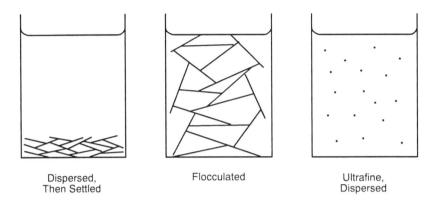

Fig. 8.1 The sedimentation height test for dispersion effectiveness.

A disadvantage of this method for practical engineering use, as compared to scientific use, is that it can not ordinarily be used with the high solids loadings of interest in many modern ceramic processes, because the slips are opaque and so viscous that they are nearly solids, except right near the point of minimum viscosity, which would have to have been known beforehand in order to do the experiments. For that reason, the next method is one of the more useful ones for realistic engineering slips.

Maximum Solids Loading At Maximum Usable Viscosity. In this procedure, a ceramic powder is ball milled for an hour with a solvent and a smaller than usual amount of dispersant, then more ceramic powder is added and ball milled for another hour. Additional increments of powder are each milled into the slip for additional hours, until the viscosity becomes too high to be usable in the process, whatever viscosity that is (typical values are given in Table 6.1). Usually the viscosity begins to rise quite steeply as the high solids loadings are approached. This phenomenon had previously been shown in Fig. 6.14, and it is visible again on any single curve of Fig. 8.2. The next step is to add more dispersant, and then repeat the above procedure, noting whether more powder can be put into a usable slip this time. A family of curves is generated, as illustrated in Fig. 8.2, and the one with the highest "solids loading" (highest volume % powder) indicates the optimum amount of dispersant. Ordinarily, the experimenter need not actually measure the viscosities except for a few right near the maximum usable value, since the other values that are far from the usable maximum can be estimated by pouring the slip. After the most effective dispersing conditions have been found, the experiment should be repeated with only one ball milling, to prevent the possibility of an error due to multiple and therefore excessive millings of the material. Even if high solids loading is not desired in some particular process, (such as making porous ceramic filters, etc.) it is helpful to know the best dispersant concentration, so that the system is less subject to random variations from lot to lot surface area changes, temperature changes, etc. (The first two tables in Chapter 10 show examples of vol. % calculations.)

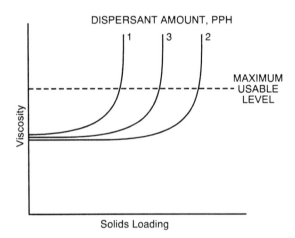

Fig. 8.2 The maximum loading test for dispersion effectiveness. (For example, the 1, 3, and 2 might be parts per hundred of dispersant.)

Other methods for estimating the maximum usable solids loading are occasionally described in the literature, an example being the addition of liquid until the porosity of the system is just barely filled. This point is called the "critical powder volume concentration (CPVC)," and it is sometimes used in the paint industry. It has been recommended for use in ceramic processing,[6] but it has not been found useful in the author's experience, because the viscosity is too high for shaping at this point.

Maximum Green Density. As discussed in Chapter 6 on colloids, the high viscosity associated with poor quality dispersion is due to agglomeration, and sometimes the whole container of viscous slip is really all one large agglomerate, although it might be rather weak. This bonding together of the particles prevents them from sliding against each other during compaction, so whatever the compaction method is (dry pressing, etc.), the packing factor in the green state is not as high as if the system had been thoroughly deagglomerated. In other words, the green density will not be maximized unless the dispersant is working properly. Other variables can also affect the green density, such as the lubrication, die design, or amount of binder. but with these held constant, an improvement in dispersion will yield higher green density, so the latter can be used as a measure of the dispersion effectiveness.

Figure 8.3 shows two methods for defining the volume during a density measurement. The weight of a green sample divided by the "bulk volume" is the "bulk density," which is a useful measurement[7a] for evaluating the degree of deagglomeration that had existed before the slip had been dried. The volume can be determined with a flat faced caliper on a simple geometric shape of green body, such as a cylinder. An average of several measurements is usually needed for accuracy. The same weight divided by the "apparent volume" shown by the heavy outline provides the "apparent density" of the green body,[7b] and a reference fluid is used to measure this so it can penetrate into the open porosity. A helium or a nitrogen type of gas pycnometer

---

[6] B. C. Matsuddy, et al., Advances in Ceram., 9 (1984) 246.

[7(a)] Anon., "Bulk Density, ASTM C373," Amer. Soc. for Testing Mat'ls., Philadelphia (1990); [(b)] Anon., "Apparent Density, ASTM E12," Amer. Soc. for Testing Mat'ls., Philadelphia (1990).

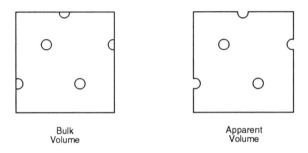

Bulk
Volume

Apparent
Volume

Fig. 8.3 Two volume measurement standards for determining density.

can be used, or a mercury porosimeter at nearly zero applied pressure. The usefulness of this density will be discussed in the chapter on binders.

Table 8.1 shows an example of the calculation of a powder packing factor from a *bulk green density,* using data from page 439 of reference 5. When the dispersion is good and other factors are also satisfactory, the packing factor of fairly fine powders should usually be well over 50% in order to sinter to full density. Therefore the green density is a good quality control tool in mass production. If the packing factor is found to fall to an unusual level (about one standard deviation below a running average), the engineer can look for problems in the deflocculation step, as a possible cause of the decrease. Also, the viscosity of the wet slip might be found to have risen at the same time as the dried green density fell. In aqueous systems, sometimes an increase in the surface area of the incoming ceramic powder might be the cause of this, in which case more dispersant can be added. Sometimes the incoming lot of dispersant has too high a molecular weight and is out of specification. In nonaqueous systems, those same two problems could cause the green density to fall, or possibly an additional problem has occurred, such as water entering the system (see discussion below).

Table 8.1 Calculation of Packing Factor from Green Density

The following had been put into the composition:
>   100 gm of alumina[††]
>   13.6 gm nonvolatile organics

Total: 113.6 gm, all materials

The bulk green density was measured as 2.5 gm/cc.
Therefore, in one cc there is 2.5 gm total material.
Of this, the weight of only the alumina is:
>   2.5 gm (100/113.6) = 2.2 gm.

The density of alumina is 4 gm/cc.
Therefore the volume of the 2.2 gm of alumina is:

$$\frac{2.2 \text{ gm}}{4 \text{ gm/cc}} = 0.55 \text{ cc.}$$

Then, $\frac{0.55 \text{ cc alumina}}{1 \text{ cc total material}} = 55\%$ of theoretical density.

(Alternatively, the organics can be burned out and the ceramic powder simply weighed, provided there is no dehydration weight loss in the powder.)

Green Density Measurement:
If the green body has a simple shape such as a cylinder, measure the size with a flat faced micrometer caliper several times and average these values. Then calculate the volume and divide that into the weight. If the green body is an odd shape, weigh (which yields W_1), then soak in melted wax, firmly wipe off any excess wax while warm, weigh again when cool (W_2), and suspend in water (W_3). Be sure to use fine wire and knock off bubbles. W_2 minus W_3 is the volume of the body, so the green density is $W_1/W_2—W_3$. (If there is no significant *open* porosity, the ASTM C914 wax method can be used [see Vol. 15.01, page 270 (1995)]. Alternatively, a mercury porosimeter can be used to measure the green body volume, by extrapolating volume measurements to zero applied pressure, or at least to a very low applied pressure.[**])

[††] Data from tape cast alumina samples [see Table 30.2 and p. 440 of ref. 5].
[**]   D. J. Shanefield, Am. Ceram. Soc. Bul., 51 (1972) 646; ASTM C 493-93.

Table 8.1 Calculation of Packing Factor, Continued

Porosity Calculation:
The weight of the organics is 2.5 x 13.6/113.6 = 0.3 gm. If the density is estimated as 1 gm/cc, then the volume is 0.3 cc. This plus the 0.55 cc of alumina is 0.85 cc total. Therefore the porosity is 15%. The organics might be sufficiently stretchable to mostly fill this volume, in which case polymer scientists would call this "free volume," rather than porosity.

Shrinkage Calculation:
As calculated in the first dagger footnote of Chapter 5, in this case the "shrinkage factor" is $0.97^{-1/3} / 0.55^{-1/3} = 0.83$. The firing shrinkage is $1 - 0.83 = 0.17$, and ceramists often express this as "17% shrinkage."

Possibly the following data will make the shrinkages more understandable. If 100% of theoretical density were achieved, the volume would be 1 cc, and the length would be 1 cm. At 97% of T.D., vol. = 1/0.97 = 1.03093 cc and L = 1.01020 cm. In the green state, at 55% of T.D., vol. = 1.81818 cc, and L = 1.22052 cm.

8.2 Commonly Used Detergents[8]

Anionic Surfactants. Figure 8.4 shows the structures of three types of surfactants which are in widespread use in both homes and factories as aids for cleaning, in other words, for use as "detergents." The negatively charged part of the anionic detergent becomes adsorbed onto dirt during washing, and this charges the dirt particles. They then electrostatically repel each other and make a fairly stable colloidal

---

[8] McCutcheon's Emulsions and Detergents," McCutcheon's Publ., Glenn Rock, NJ (1993) [technical data and supplier addresses for hundreds of surfactants]; R. B. McCay, "Technical Applications of Dispersants," Marcel Dekker, New York (1994).

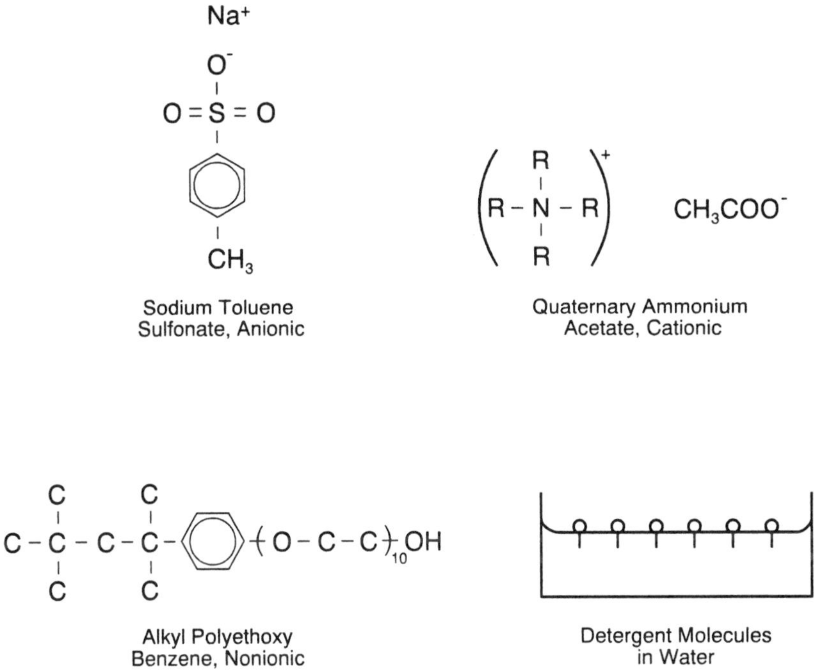

Na+

O⁻
|
O = S = O
|

CH₃

Sodium Toluene
Sulfonate, Anionic

$$\left(R - \underset{R}{\overset{R}{\mid}} N - R \right)^{+} \quad CH_3COO^{-}$$

Quaternary Ammonium
Acetate, Cationic

C   C
| |
C - C - C - C -〈 〉-(O - C - C)₁₀OH
| |
C   C

Alkyl Polyethoxy
Benzene, Nonionic

Detergent Molecules
in Water

Fig. 8.4 Common detergents. A good cationic quat salt for ceramic use has three R groups consisting of ethyls, with the fourth being a polyethylene glycol having n=12.

suspension. The initial removal of the dirt had to be done mostly by mechanical action, and the detergent mainly prevents redeposition to the clothing or dishes, etc. This is analogous to the action of a dispersant during the ball mill deagglomeration in ceramic processing, where aggressive mechanical action is needed, and in fact some detergents can be used as dispersants for making ceramics. (See also literature listed under "Surfactants" in Appendix III.)

Much of the dirt that is to be washed away consists of clay-like earthy powder, similar to ceramic powders. However, it is most often oily. Therefore most washing detergents have a nonpolar or slightly polar part of the molecule that makes van der Waals bonds to the nonpolar or slightly polar oil on the dirt. Because of the H-bonds in water, this nonpolar end of the molecule gets pushed away from the water and towards anything else, such as an air-water surface, or a solid-water surface. The other end of a typical anionic detergent molecule is highly ionic, and it stays in the water, easily forming H-bonds. Thus these molecules tend to line up on the water surface, with heads out and tails in. They are pushed toward anything nonpolar and will generally stick to it. By van der Waals bonding, they can stick to ionic dirt, also, since that kind of bond can form to some degree on any known material. However, they bond better to oil than to ceramic powder, because the latter bonds more to the water itself. Some of the detergent can stay inside the bulk of the water, because it is somewhat soluble in water. Since each molecule has one or more negatively charged ionic ends, whatever it sticks to becomes negatively charged.

Another consequence of the segregation onto the water surface is that the surface tension of the liquid becomes that of the surfactant, since this is now the new surface (see the diagram of a water container in the figure). This new surface consists of mostly oily, nonpolar half-molecules which have low surface tension. The water solution of surfactant will wet oil or wax, while pure water would ball up and not wet them, because of its high surface tension. If a ceramist gets a new lot of clay which has been stored near an oil refinery, it might be microscopically coated with oil and impossible to mix uniformly with pure water to make a slip, because the water refuses to wet the contaminated clay. In that case, a little bit of surfactant added to the water makes a complete change in the wettability and allows smooth mixing.

Household soap bars consist mostly of sodium stearate, made by reacting sodium hydroxide with glyceryl tristearate ester, which is the major component of animal fat. This is the reverse of the reaction for making the ester shown in Fig. 3.14. Excess sodium hydroxide is present during soap manufacture, so the stearic acid that forms when the ester is decomposed immediately reacts to form the sodium stearate salt. Since stearic acid is a weak Bronsted acid, it is not very soluble in water, because it does not ionize sufficiently to dissolve. However, an excess of alkali (base) drives the ionization forward by "mass action," in which a large amount of starting material and small amount of product tend to drive any reaction forward. Stearic acid will dissolve in ammonium hydroxide also, making ammonium stearate, which is used in ceramics as a lubricant for dry pressing. The ammonium salt would not leave any oxide residue after burnout, as the sodium salt would.

Another reason to have some excess alkali present is that it can make the surface of a natural oil into its own detergent, if that oil is an ester. Thus many oily materials can become colloidal sols if they are broken down mechanically to small particles in the presence of a strongly basic solution. In addition, any clay or similar powder that is present can become negatively charged by its own adsorbed water becoming ionized in base, as was shown in Fig. 6.4. For these reasons, detergents are made moderately alkaline. If the ceramist wishes to make an emulsion (that is, an oily colloidal sol) of wax to use as a new lubricant for extrusion, etc., having some excess ammonium hydroxide present is often helpful.

Sodium toluene sulfonate (see figure) is the main component of dishwasher detergent. It is inexpensive, and the π bonds help the aromatic ring make strong van der Waals bonds to oily materials. The sulfonic acid group is a medium strength Bronsted acid, as compared to very weak stearic acid, so it dissolves in water without needing to be too alkaline. Ceramists ordinarily do not like to use sulfonates or sulfates, because they might leave uncontrolled sulfur compounds behind after burnout. If they do evaporate out, sulfur compounds might make such strong bonds to the ceramic and its inorganic additives (such as sintering aids) that some of the inorganics are also evaporated along with the sulfur, changing the composition uncontrollably. This is the same reason that we do not usually put chlorides into an advanced ceramic green body: they might cause something to volatilize out during firing. However, it is debatable whether or not this is an exaggerated concern, since there would not be much sulfur or chlorine present if it is only a minor additive.

Cationic Surfactants. Surfactants can be made with the large, readily adsorbed part of the molecule being positively charged, instead of negatively, and an example of such a cationic detergent is shown in Fig. 8.4. In most of these cationics, all four hydrogen atoms in the ammonium ion shown previously in Fig. 2.1 have been replaced by organic radicals, and an example is tetramethylammonium hydroxide (not shown in the figure), which is a Bronsted base just like ordinary ammonium hydroxide is. A chemist can start with uncharged, nonionizable trimethylamine, $(CH_3)_3N$ (compare ethylamine, Fig. 3.13), and then an extra methyl group can be added, giving the whole structure a plus charge. After this is done, the positive charge is distributed by quantum mechanical resonance thoughout the whole positive ion, and one can not tell which were the original three methyls. These cations are sometimes referred to as "quaternary ammonium" ions, or simply "quats."

A much easier alternative way to make an organic base is to simply put trimethylamine into water, in which case it will spontaneously form trimethylammonium hydroxide, $(CH_3)_3NH^+ OH^-$ which is analogous to adding ammonia gas to water and forming the better known but similar $NH_4^+ OH^-$ base. The more organic radicals attached to the nitrogen, the stronger the base, so ammonium hydroxide is only a moderately strong base, trimethylammonium hydroxide is slightly stronger, and tetramethylammonium hydroxide is much stronger, being similar in its degree of ionization (essentially 100%) to sodium hydroxide. Any of these can be used where a strong base is needed but it must be completely burnable. Organic groups other than methyls can be put on the nitrogen, to optimize adsorption to oil, etc.

A tetra-alkyl ammonium base can be reacted with an acid to form a salt, which would be the cationic detergent shown in Fig. 8.4. The large, somewhat polarizable and highly charged cation tends to become adsorbed onto oil droplets, giving them positive charges so they will repel each other and remain a fairly stable sol. They do not bond very well to ceramic powder, because the H-bonding capability of the nitrogen has already been taken up by the extra organic groups, but they do bond to some extent by van der Waals, and they can be used as dispersants. Mainly they are used for emulsifying oils and waxes in water, to be used as floor polishes, cosmetics, and the like. They are used in ceramic processing to make sols of lubricants and also any binders which are insoluble but emulsifiable, like wax or PMMA acrylate. The acetate ion shown in the figure could be substituted by sulfate, chloride, phosphate, or other acid ions, but something burnable like acetate is more desirable for ceramic applications.

In strongly acidic slips, where the alumina and other ceramic powders develop their own positive charges from ionization of adsorbed water (Fig. 6.4), soap or other anionics usually will not work, because they add a negative charge which conflicts with the powder's own positive charge in acid (Fig. 6.8). However, cationic quats, since they add more positive charge to the powder, do work well in acid. This might be a useful thing to remember when one component such as an already emulsified acrylate binder is working satisfactorily in a ceramic slip, and another material such as a lubricant must be emulsified and then added to the composition. If the first one is acidic, it might not be compatible with anything else except another acidic system. Aside from the acidity factor, it is not a good idea to mix cationics with anionics even in neutral water, because their charges would tend to neutralize each other.

With nonpolar oils, which do not have adsorbed OH groups on the surface and therefore do not ionize, cationics will also work in alkali, and the high pH enhances the overall action. Actually, any of these surfactants can work as dispersants at the "wrong" pH, by operating mainly through steric hindrance, and not by charge repulsion, but in water the charge effect is usually more effective, because a very strong effect is needed to compete with H-bonding.

For emulsifying high concentrations of waxes into water, as in automobile polish, cationic quats seem to work better than the anionic detergents do. A procedure[9b] for emulsifying wax in water using cationics is the following:

(1) Melt the wax (or dissolve the stearic acid in hot ethanol).

(2) Pour the melted wax (or solution) into stirred, nearly boiling water that contains 1% of quat salt.*

(3) Although this step is not always necessary, a mechanical "emulsifier" that breaks down the

---

9 S. M. Marschner, Ph.D. Thesis, Rutgers Univ., (May 1991) pages 42, 51, 59 [available from UMI, Ann Arbor, Michigan]. (See also page 277 of this book.)

* Product CC-55, from Witco, Stamford, CT.

oily particles to submicron sizes can then be used.
Laboratory supply companies sell small units,
sometimes listed as "homogenizers."

Cationic quats are such powerful wetting agents, even at neutral pH, that they can make bacteria wettable. The bacteria, like some insects, must be water repellant in order to prevent osmosis from pumping too much water into them, and when they become wettable, they die. Therefore, quats are used as disinfectants, especially in throat lozenges for treating colds, eye drops, antibiotic medicines, etc., where the pH has to be about 6. They are popular as "biocides" for ceramics, where they can kill algae that would otherwise grow in slip storage tanks. Mercury compounds were previously used as biocides, but human toxicity now precludes their use.

Many food products such as milk and mayonnaise are emulsions. In the first of these the oil is dispersed in the water, that is, the water is the only continuous phase. (Some of these concepts are explained more completely in Appendix II.) In mayonnaise, however, the oil is the continuous phase. Working with equal amounts of oil and water, one can sometimes choose which will be the continuous phase by selecting the proper surfactant. A scale of polarities similar to the solubility parameter scales is available for making emulsions. It is called the hydrophilic-lipophilic balance, or HLB system,[10] and it might be useful for specialized applications such as performing emulsion polymerization to make reusable porous plastic molds for slip casting.

Nonionic Surfactants. One of the most common nonionics is shown in the figure. Although it is not polar enough to ionize, one end is slightly more polar than the other, so it migrates to the liquid surface. It has a powerful effect in lowering the surface tension of water to make it wet oily surfaces, and it is used widely in such applications as photographic developers, electroplating solutions, etc., where it does not interfere with other necessary functions since it is not ionic. Being nonionic, it does not irritate human eyes, so this type of surfactant is used in hair shampoo. Another feature, for better or worse, is that it tends to stabilize bubbles by a delicate balance of effects on surface

---

[10] M. J. Rosen, "Surfactants," J. Wiley, New York (1978); Anon., "The HLB System," ICI Americas Inc., Wilmington, Del. (1980).

tension, so some solutions of nonionics tend to have powerful foaming effects. This is also desirable in some hair shampoos, but not in most ceramic or other technical applications.[11]

Nonionics do not interfere strongly with other surfactants, so they often are added to the other two types to enhance wetting, provide foam, etc. Being not very polar and quite inert chemically, they are not thoroughly removed from sewage water by the usual waste treatment operations, so large quantities have accumulated in the environment. The compound shown in the figure has been detected in oceans and rivers all over the world and in many life forms, and it is certainly too late for it to be removed by any imaginable environmental cleanup. While this compound was thought to be biologically inert, some recent evidence indicates that it or related nonionics might interfere with animal sexual functions (see reference 40 in Chapter 7). If we are

---

[11] In the 1980s, the biggest printed circuit factory in the world was in Richmond, VA, owned by AT&T. A serious problem of poor wetting developed in a copper plating tank, resulting in the rejection of expensive products. This factory was only marginally profitable ordinarily, and with the low yield it became a huge money loss for AT&T. The plant manager had been told that his factory would be shut down, *permanently,* if this was not fixed, but the workers had not yet been told. The nonionic detergent shown in Fig. 8.4 was tried by the author in a laboratory size tank, and it solved the problem. Stirring was typically done in plating tanks by having some air bubbling upwards, so there was a small amount of foam. The author and his loyal technician went to Richmond and added the detergent, scaling up the amount in proportion to the tank volumes, lab versus production. It turned out that the detergent was not very soluble and mostly collected on the top *surface* of the plating solution, which called for a scale-up proportional to the *square* of lab versus producton tank widths, not the cubic scale-up that was used. Excess nonionic detergent caused the entire tankload of poisonous copper solution to foam up over the side of the huge tank, like a green monster in a horror movie. Fortunately, it all went down into the safety floor drain. Unfortunately, the piping had mistakenly been directed to the *sanitary* waste treatment plant instead of its chemical counterpart, and the copper killed all the beneficial bacteria in the sand filters. With no operable waste treatment, the factory had to be closed and the workers furloughed without pay. Upon returning to the research laboratory, the author's loyal technician told the lab manager, "Dan always said 'If you have to fail, fail big.' And he *did* — he got us kicked out of the biggest printed circuit plant in the world!" During a frantic week of experiments, the correct amount of detergent was used, the problem was solved, and the plant is still running. However, the factory workers only seemed to care about the loss of a week's pay, and there are some in Richmond that the author still would not like to meet face to face in a dark alley.

lucky, this will turn out to be another false alarm. If we are not lucky......

Since nonionics can not impart an electrostatic charge on ceramic particles, they can only work as dispersants by steric hindrance.[12a] In aqueous systems, this is not a strong enough effect to overcome the competing tendency of H-bonding to form agglomerates, so nonionics are only occasionally used for that purpose. However, in nonaqueous systems the typical "polyether" nonionics are used as dispersants.[12b]

8.3 Inorganic Surfactants

The top of Fig. 8.5 shows compounds of six chemical elements, arranged in the same order that appears in the periodic table:

Group 3	Group 4	Group 5	Group 6	Group 7
Boron	*Carbon*	*Nitrogen*	*Oxygen*	Fluorine
Aluminum	*Silicon*	*Phosphorus*	*Sulfur*	Chlorine.

The compounds in the figure are the oxy-acids of those elements, and there are similarities based on the increasing electronegativities and valences going toward the right hand side of the periodic table. The bottom three can form dimers or polymers, and the polymer anions will adsorb onto ceramic powders or wax particles more than small ions would, causing large electrostatic repelling charges to develop. In addition to acting directly as dispersants, these compounds "complex" or bond strongly to the multiply charged cations such as Ca^{++} and Mg^{++} that are present in "hard" water and which can collapse the double layer of otherwise stable colloids, causing the sols to flocculate. Therefore the polymers shown in the figure indirectly help to stabilize sols, and they are called "builders" when added to detergents to build up the overall cleaning action. Boron, in Group 3, forms similar compounds, but for simplicity it is not shown in the figure.

12(a) Gum arabic, a fairly good dispersant for ceramics, was shown to operate entirely by *steric hindrance* in water, not by *repulsion*, since there was no measurable elctrophoretic mobility: see E. M. Vogel, Am.Cer.Soc.Bul., 58 (1979) 453. (b) A. Roosen, Ceram. Trans., 1B (1988) 675 [see particularly table 3].

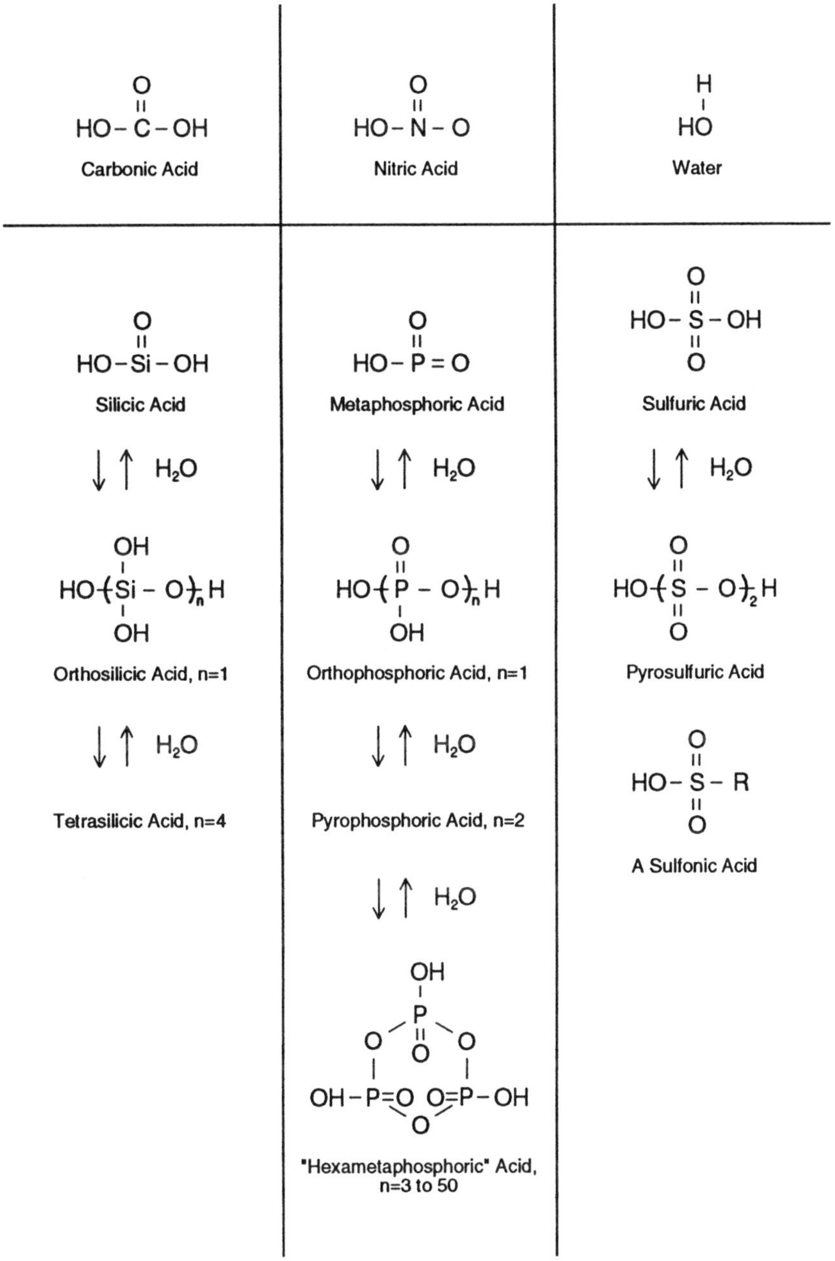

$$
\begin{array}{ccc}
\overset{\displaystyle O}{\underset{\displaystyle}{\parallel}} & \overset{\displaystyle O}{\underset{\displaystyle}{\parallel}} & \overset{\displaystyle H}{\underset{\displaystyle}{|}} \\
HO-C-OH & HO-N-O & HO \\
\text{Carbonic Acid} & \text{Nitric Acid} & \text{Water}
\end{array}
$$

Carbonic Acid — HO–C–OH with O double bonded to C

Nitric Acid — HO–N–O with O double bonded to N

Water — HO with H

Silicic Acid — HO–Si–OH with O double bonded to Si

Metaphosphoric Acid — HO–P=O with O double bonded to P

Sulfuric Acid — HO–S–OH with O above and O below S

↓↑ H_2O

Orthosilicic Acid, n=1 — HO$-$(Si$-$O)$_n$H with OH above and OH below Si

Orthophosphoric Acid, n=1 — HO$-$(P$-$O)$_n$H with O above (double bond) and OH below P

Pyrosulfuric Acid — HO$-$(S$-$O)$_2$H with O above and O below S

↓↑ H_2O

Tetrasilicic Acid, n=4

Pyrophosphoric Acid, n=2

A Sulfonic Acid — HO$-$S$-$R with O above and O below S

↓↑ H_2O

"Hexametaphosphoric" Acid, n=3 to 50

Fig. 8.5 Some inorganic dispersants. ("Silicic" acid is sometimes called "metasilicic.")

Sodium tetrasilicate, sometimes called "water glass," is used in large quantities as a dispersant for clay. Being a weak acid, it hydrolyzes (reacts with water) to some degree in aqueous solution, forming the strong base sodium hydroxide and the much weaker tetrasilicic acid, so the net result is that the solution is strongly basic. This basicity causes the clay to become negatively charged because of the ionization of its own adsorbed OH groups, and the tetrameric silicate ions also get adsorbed and add several more negative charges. To some degree, the polymerization (organic chemists would call it oligomerization) changes with time and temperature, causing a slight drift in properties, but it is satisfactory for most purposes, and there are many detailed reports of its use.[13] After the ceramic is fired, residues of about 1% sodium oxide and silica do not significantly change the properties, because they are already present in clay.

The polyphosphates similarly polymerize, and they also change with time and temperature like the silicates do. They make good dispersants,[14] as long as the inorganic residues that are present after firing do not cause any problems with the particular ceramic being made. The salts called "*hexa*metaphosphates" actually might contain other than *six* monomers (Fig. 8.5 shows a "*tri-*" acid). This depends on the temperature, concentration, and past treatment history (the polymerization and depolymerization reactions are quite slow to reach equilibrium, so the n value is "path dependent"). One particular form with n approximately 21 has excellent properties when used with the sulfonates as an auxiliary detergent or builder, as described above. Because phosphorus can be a nutrient for algae, many local governments have limited the amount allowed to be put into detergents, so only a small percentage is used in commercial soap flakes, along with larger percentages of the somewhat less effective builders, sodium tetrasilicate and sodium tetraborate. The author and a student[15] were able to disperse 68 volume % of Alcoa A-16SG alumina in an aqueous solution of 3 weight % sodium "hexametaphosphate" (a salt with n

---

[13] G. W. Phelps and M. G. McLaren, page 211 in "Ceramic Processing Before Firing," G. Y. Onoda and L. L. Hench, eds, J. Wiley, New York (1978) [Fig. 17.4]; Anon., Am. Ceram. Soc. Bul., 71 (1992) 185; W. Kohut, Am. Ceram. Soc. Bul., 71 (1992) 947; A. B. Corradi, et al., J. Am. Ceram. Soc., 77 (1994) 509.
[14] J. Faisson and R. A. Haber, Ceram. Eng. Sci. Proc., 12 (1991) 106.
[15] Siddhartha Basu and D. J. Shanefield, unpublished work.

approximately 15), using a ball mill. (However, possibly the fines "dissolved," because of the high pH, and this was difficult to determine, or even to define.)

Although the sulfuric acid dimers do not seem to adsorb strongly onto powders, one of the OH groups in sulfuric acid can be replaced by an organic radical, and the resulting compound is the sulfonic acid shown in the figure. The organic sulfonates do adsorb strongly onto many powders and oils and are therefore very useful surfactants. Sodium lignosulfonate (Fig. 3.25) and sodium toluene sulfonate (Fig. 8.5) are compounds of this type. One more oxygen can also be put onto the sulfur, making organic sulfates, and these are also used as detergents to some extent, although they are more expensive.

8.4 Organic Deflocculants For Ceramics

Aqueous Systems. A floc, as defined in Appendix II, is a loose agglomerate with an open sort of structure. In some instances the solid particles are actually touching each other, as has been shown with graphite particles flocced in oil when they conduct electricity.[16] In other types of flocs the particles are separated by very thin layers of liquid, and graphite in those cases does not conduct, although the viscosities in both cases might be similar. In neutral water, flat clay particles usually have natural charges as shown in Fig. 6.1, and this can cause them to floc in a "house of cards" arrangement,[17] similar to cardboard playing cards arranged edge to face in a delicately balanced house-like structure (Fig. 8.1).

As discussed in Chapter 6, lowering or raising the pH can overwhelm the natural plus and minus charges on clay and force a single strong charge to form on the surface, so that mechanical deflocculation will then lead to a repulsion-stabilized sol.

---

[16] G. D. Parfitt, "Dispersion of Powders in Liquids," Elsevier, N. Y (1981) 315.
[17](a) H. van Olphen, "Clay Colloid Chemisty, " J. Wiley, New York (1977) 167;
(b) J. S. Reed, "Principles of Ceramics Processing," J. Wiley, New York (1995) 175 [see Fig. 11.1, left-hand diagram].

Alternatively, the negative or positive ions of an organic dispersant can adsorb onto the powder, forcing even stronger new charges to form, and large ions usually adsorb more readily than small ones. Also mentioned in Chapter 6 was the fact that negative dispersant ions will usually adsorb more if the solution is slightly acid, making the powder positively charged and thus attracting the ions. There have been many reports of this,[18] and often there is an optimum pH for achieving the maximum charge, which is only slightly acidic. Too strong an acid can force the organic dispersant molecule to become less ionized, by the principle of *mass action*.

The opposite charges would be present if the dispersant is a cation such as polyethyleneimine.[19] The imine group (Fig. 3.13) when put into water becomes similar to an ammonium cation, or to the cationic quat ion in Fig. 8.5, and then a slightly alkaline solution would increase the adsorption because of a negative powder attracting these positive ions.

Several factors that lead to a large amount of the dispersant becoming adsorbed have been brought up previously in this chapter and in Chapter 6. It should be noted that many powders such as silicon nitride have both acid and base sites,[20] so a *block copolymer* dispersant having both (called "zwitterions" if both parts are ionized) can be doubly attracted to the powder.[21] Sometimes a copolymer dispersant with both polar and nonpolar parts of the molecule will be a more effective dispersant than a simpler one would be, because the polar end adsorbs to the particle by enhanced van der Waals attraction, and the nonpolar end sticks out into the solvent and mechanically fends off other particles.[22]

---

[18] A. Foissy, J. Colloid and Interface Sci., 96 (1983) 275; J. Cesarano and I. A. Aksay, J. Am. Ceram. Soc., 71 (1988) 1062; Z. C. Chen, T. A. Ring, and J. Lemaitre, Ceram. Trans., 22 (1991) 257; K. Nagata, Ceramic Trans. , 22 (1991) 335 (see Fig.2); K. Nagata, Ceramic Trans. , 26 (1991) 205 (see Fig.3).

[19] A. Bleier, et al., Colloids Surf., 1 (1980) 407.

[20] L. Bergstrom, et al., J. Am. Ceram. Soc., 72 (1989) 103; S. G. Malghan, et al., Ceram. Trans., 26 (1992) 38; J.-F. Wang, R. E. Riman, and D. J. Shanefield, Ceram. Trans., 26 (1992) 240.

[21] M. Liphard and W. von Rybinski, Progr. Colloid & Polymer Sci., 77 (1988) 158.

[22] (a) R. J. Hunter, "Foundations of Colloid Sci.," Clarendon Press, Oxford (1987) p. 489; (b) J. F.Stansfield, U.S. Pat. 3,996,059 (1976); (c) A. Topham, U.S. Pat.

The optimum molecular weight of the dispersant varies according to several factors, but it seems to be close to about 1,000 for many examples of ceramic systems. If the dispersant is something like a polyacrylate salt (Fig. 3.16), there are repeating ionic groups along the chain, so the higher the m.w., the more charge could be added to the powder particle. However, a dispersant made of this material but having too low a molecular weight will not be adding enough charge to be effective.[23] On the other hand, too high a molecular weight can also lead to various conditions that are less than optimal. The long molecule can snake around in the "loop-train" configuration[24] shown in Fig. 8.6, where some of its effectiveness might be wasted. Ideally, it

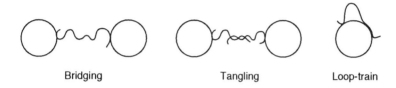

Bridging   Tangling   Loop-train

Fig. 8.6 Some possible configurations for long adsorbate molecules.

sticks out radially, as in the alkaline solution example shown in Fig. 6.2, tending to keep other particles away even if they approach with higher than usual velocities. Actually, a little bit of loop-train configuration can make stronger bonds to the powder, but too much is not effective.

Another problem with excess chain length is "bridging," whereby one end of a long molecule adsorbs onto a particle, and the other end adsorbs onto another particle (Fig. 8.6), causing the flocculation that

4,224,212 (1980); see also "imine" on pages 13, 244, and 287. [Material is similar to KD-3 polyimine ester copolymer, from ICI Co., Wilmington, DE.]

[23] L. Romo, et al., Disc. Faraday Soc., 42 (1966) 232; T. Allen and R. M. Patel, J. Colloid and Interface Sci., 37 (1971) 595; A. Doroszkowski, et al., Faraday Discuss'n of Chem. Soc., 65 (1978) 252; F. F. Lange, et al., J. Am. Ceram. Soc., 77 (1994) 922.

[24] B. M. Moudgil, et al., Advances in Ceram., 21 (1987) 483 [Fig. 1].

the ceramist is usually trying to avoid.[25] This is done purposely in making *flocculants,* for promoting porous agglomerates to be used as filters in waste treatment, etc. Long polyacrylates and polyacrylamides (m.w. about a million) are used with aluminum hydroxide flocs as one of the main methods for purifying water, both incoming drinking water and outgoing sewage. Bacteria sols and even virus solutions are chemically adsorbed and/or mechanically filtered out by this method. Water treatment and latex paint are the biggest uses of polyacrylates, and the ceramics dispersant market is small by comparison, although it can be profitable for particularly effective products. (For the definition of latex, see Appendix II.) In addition to *bridging,* very long molecules from two particles will tend to tangle (called "interpenetration" in the older literature), which can also cause flocculation.[26] Probably for both of these reasons, examples have been reported[27] where a polyacrylate m.w. of about 1,800 was more effective than higher molecular weights in achieving a low viscosity.

Long organic molecules are known to be broken into smaller fragments when they are dissolved and the solution is stirred at very high shear rates. When polymer chemists measure m.w. using chromatography with too fast a pumping rate, the m.w. appears to be lower than when slower pumping is used. Once done, the effect is permanent, that is, repeating it at a slower rate will still give the low m.w. reading. From this it has been concluded that long chain molecules can be rather fragile. Ceramists have observed that ammonium polyacrylate dispersant of several thousand m.w., when ball milled for many hours, appears to change its properties. For this reason, one supplier offers a low m.w. polyacrylate (about 6,000) for use whenever the milling time is more than an hour, and it is claimed that the properties remain constant.[28]

---

[25] R. J. Hunter, "Found'ns of Colloid Sci.," Clarendon Press, Oxford (1987) p. 489; R. G. Horn, J. Am. Ceram. Soc., 73 (1990) 1117 [see particularly page 1126].

[26] D. J. Shaw, "Introduction to Colloid Chemistry," Butterworths, N.Y. (1980) p.208; I. Sushumna and E. Ruckenstein, J. Mat'ls. Res. 7 (1992) 2884 (see especially page 2888].

[27] J. Cesarano and I. A. Aksay, J. Am. Ceram. Soc., 71 (1988) 1062 [at 50 volume % solids loading of a half micron powder, see Fig.5].

[28] Technical specification sheets and sales literature from R. T. Vanderbilt Co., Norwalk, CT (1994), regarding Darvan 821A" (low m.w.) product, versus the older Darvan C (high m.w.) product.

At the ultimately high solids levels, where powder particles are nearly in contact instead of repelling each other from a greater distance, the thicker the adsorbed dispersant layer that coats each particle, the less actual powder can be fitted into a closest packed volume. This was demonstrated with high resolution electron microscopy, where the coating could be seen to prevent particles from quite touching.[29] Possibly the use of low m.w. dispersants in such cases would allow closer packing, or possibly it is just the total weight of dispersant adsorbed that would the critical factor.

It has been found in some systems that branched chains are more effective than straight chains,[30] especially at very high solids loading where the molecular shape can be critical. Fairly long sidechains (around C_{18}) were reported[31] to be more effective than C6 sidechains attached to the polyacrylate backbone, where the packing factor was so high that the dispersant could really be called a "lubricant." Also at very high solids, some newer dispersant systems attempt to go beyond just depending on the DLVO concept of electrostatic repulsion, and they use high salt concentrations to make "non-DLVO," densely packed pastes[31, 32] (see also reference 32 in Chapter 6). Adsorbed water, which is always present to some extent in these systems, is depended upon to provide some of the lubrication, particularly during *pressure filtration.* (described in refs. 31 and 32 of this chapter) or *centrifugation* (described in ref. 72b in this chapter).

Even in ordinary aqueous DLVO systems, in addition to charge repulsion, the *steric hindrance* from adsorbed dispersant molecules normally plays a part in providing good dispersion,[12a] and this is additionally true at the ultimate high loadings achievable. In other words, many dispersants that are liquids or soft solids can act as lubricants at very high solids loading, adding to the charge repulsion effect. This is analogous to having a rubber bumper at the end of the suspension spring in an automobile. When encountering particularly large bumps, if the spring becomes completely compressed and

[29] I. A. Aksay, Ceram. Int'l., 17 (1991) 267 [see particularly page 272].

[30] A. Doroszkowski, et al., Faraday Discuss'n of Chem. Soc. , 65 (1978) 252;

T. K. Yin and I. A. Aksay, Ceram. Trans., 1B (1988) 654;

[31] I. A. Aksay, et al., Ceram. Trans., 1 (1988) 654

[32] W. A. Ducker, Z. Xu, D. R. Clarke, and J. N. Israelachvili, J. Am. Ceram. Soc., 77 (1994) 437; J. C. Chang, F. F. Lange, and D. S. Pearson, J. Am. Ceram. Soc., 77 (1994) 19.

"bottoms out," the rubber bumper can still take up the rest of the shock. Most aqueous systems appear to make use of both steric hindrance and ionic repulsion, to various degrees.

Many different organic dispersants are used with water-based slip systems in commercial practice. In addition to polyacrylates, good dispersion results have been reported with sodium CMC,[33] lignosulfonates,[34] toluene sulfonate and naphthalene sulfonate,[35] PEG,[36] and salts of tannic acid.[37] As was mentioned briefly in Chapter 3, some of the natural materials act as dispersants and/or binders, examples being gums, alginates, and the humic acid in some clays. The chemical structures of many of the organic deflocculants used in ceramic processing have been assembled in convenient charts.[38]

While the binders can act as dispersants, the dispersants can act as binders to some degree, especially if the m.w. is large. If a range of molecular weights is present, the smaller ones might act mostly as dispersants, the oligomers might act as plasticizers, and the largest ones might have some binding action. At least one commercial additive makes use of this principle.

Solids loadings of more than 60 volume % for half micron powder have been reported in references 6 and 7 of Chapter 7 and elsewhere. (If the dispersant is the salt of a weak acid, this is usually done at a slightly alkaline pH in order to force it to ionize.) These high solids loadings are desirable for the tape casting of fairly thick ceramics, as mentioned in the discussion on evaporation rates in section 7.2. To accomplish this, however, the system must be capable of achieving high solids with some binder also being present. A serious limitation on the maximum solids can occur when large amounts of the same general type of molecule are used as both the dispersant and the binder. There is so much polymer present that "depletion flocculation" can cause a floc or gel to form. One theory attempting to explain this phenomenon is based on osmotic pressures, wherein most of the

---

[33] A. J. Ruys, et al., Am. Ceram. Soc. Bul., 69 (1990) 828.

[34] J. C. Le Bell, et al., J. Colloid & Interface Sci., 55 (1976) 60.

[35] E. Carlstrom, et al., Cer. Trans. 2(1989) 175.

[36] C. W. A. Bromley, Colloid Surf., 17 (1986) 1.

[37] H. van Olphen, "Clay Colloid Chemisty, " J. Wiley, New York (1977) 167.

[38] W. R. Cannon, et al., Advances in Ceram., 26 (1989) 525; R. Moreno, Am. Ceram. Soc. Bul., 71 (1992) 1521.

polymer remains in solution and is not adsorbed.[39] It is difficult to use this theory to predict whether the polymer will cause gelation or the opposite effect, "depletion stabilization," and in general, at least one researcher[40] in this field of study has stated that the depletion theories are "as yet not understood." (Bridging and tangling due to excessive total polymer might be a better explanation for high viscosity than the somewhat paradoxical osmotic theories, or possibly the emulsion type of binder is destabilized* by the dispersant ions.)

An example of the depletion flocculation problem is as follows. High solids loadings of fine alumina powders in water can be obtained with ball milling. However, gelation occurs during the later addition of either soluble or emulsified acrylic binder, unless the solids loading is lowered to around 40 vol. % by the use of more water, at pH≈7. The three component diagram in Fig. 8.7 illustrates the problem, which was reported by Nagata.[41] A similar limitation on solids loading in the presence of binder was also reported by Yasrebi.[39b]

However, aqueous solids loadings of 71 vol.% multisize alumina powders before adding binder, and 62 vol.% after binder additions, were achieved at Rutgers University,[39d] using 0.5 p.p.h. of NH_4 polyacrylate dispersant (m.w. = 1,800) at pH = 8.5, plus 2 p.p.h. of a highly optimized acrylate binder system (Duramax® 1001 emulsion, from Rohm & Haas Co., Philadelphia, PA).

For some applications, the dispersant does not have to be a soluble material. Instead, it can be an ultrafine powder such as certain types of magnesium aluminosilicate clays. An example is the commercial material[28] Veegum T®. In older literature, this was called a "protective colloid," but the words sometimes have other meanings in paint and ink technologies. This idea is sometimes used in modern ceramics with synthetic powders,[42] although not often because of difficulty with the reproducibility of ultrafine synthetic materials.

39(a) R. J. Hunter, "Foundations of Colloid Sci.," Clarendon Press, Oxford, UK (1987) 483; (b) M. Yasrebi, Ph.D. Thesis, Univ. of Washington (1988) [available from UMI, Ann Arbor, MI]; (c) S. J. Patel and M. Tirrell, Ann. Rev. Phys. Chem., 40 (1989) 597; (d) D. J. Shanefield and T. Suwannasiri, to be published.
* Rohm & Haas Co. suggests Triton X-405 to stabilize emulsions in such cases.
40 R. Moreno, Am. Ceram. Soc. Bul., 71 (1992) 1521 [see particularly p. 1527].
41 K. Nagata, J. Ceram. Soc. Japan, 100 (1992) 1271, in Japanese [see Fig. 6].
42 A. Garg and E. Matijevic, Langmuir, 4 (1988) 38; E. Liden, M. Persson, E. Carlstrom, and R. Carlsson, J. Am. Ceram. Soc., 74 (1991) 1335.

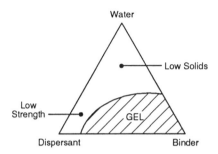

Fig. 8.7 Gelation range sometimes observed at high solids loadings when binder is added. An example might be *tert*-butyl alcohol as the dispersant (a liquid, in this case), and polyacrylic acid as the binder.

Nonaqueous Systems. In nonpolar solvents such as toluene, for practical purposes the amount of ionization possible is insufficient to stabilize dispersions, and electrophoresis shows close to zero mobility of dispersed powders. Therefore steric hindrance is the dominant stabilizing mechanism by which dispersants operate in very nonpolar systems.[43] For example, alumina dispersed with menhaden fish oil in a mixture of trichloroethylene and ethanol was not observed to move in a strong electric field. The menhaden oil does indeed act as an excellent nonaqueous dispersant, and therefore the steric theory was used to explain nonaqueous dispersion in ceramic slips.[44] (This was published about 20 years ago, and it was probably the first use of the steric theory in ceramics.) Steric hindrance was invoked essentially by default, since it would be difficult to investigate the steric effect directly. In some literature on this subject,[39] the term "steric stabilization" is used with the same meaning as the term "steric hindrance" has here.

If alcohol or other somewhat polar liquids are present, some ionization is possible, depending on the overall degree of polarity of

J. Am. Ceram. Soc., 74 (1991) 1335; E. Liden, Ph.D. Thesis, Chalmers University of Technology, Goteborg, Sweden (1994) 23 and 51.

[43] R. G. Horn, J. Am. Ceram. Soc., 73 (1990) 1117 [theory]; S. S. Patel, Annual. Rev. Phys. Chem., 40 (1989) 597 [experiments].

[44] D. J. Shanefield and R. E. Mistler, Am. Ceram. Soc. Bul., 53 (1974) 416.

the system. For example, methyl ethyl ketone (MEK, or 2-butanone) plus ethanol has been found to be polar enough for some dispersants to exhibit electrophoretic mobility of powder particles,[45] so evidently both electrostatic repulsion and steric hindrance can each contribute to stability in some nonaqueous slips. In most such systems, the ionic conductivity is relatively low, as can be estimated simply by using an ohmmeter, and steric hindrance seems to be the main stabilization mechanism, even when up to about 40% alcohol is used. However, one might make use of the observations in reference 45 for further optimization.

For any material to operate as a steric hindrance dispersant, it must be:

(1) *adsorbed* enough to cover much of the powder surface,

(2) *stick out* into the solvent enough to fend off other particles, and

(3) be *soft* enough so other particles can not effectively adhere to it.

The third point in the list above is possibly not obvious, and it is entirely the author's opinion, not generally discussed in other literature. Supporting discussion follows.

Some authors[39] state that the adsorbed dispersant must make a thick enough layer so that the particles are held far enough apart for the van der Waals attractions to be very much weakened. This appears to be correct in itself, but a question should be asked "What about the attractions of the coatings themselves to each other?" It seems logical that two particles that are well coated with a dispersant which is a moderately *strong* material can always stick to each other by van der Waals forces upon contact, just like the bare particles themselves would stick. In other words, the attraction between the underlying particles might be diminished by distance, but the coatings that are responsible for this can attract each other. This adhesion force can occur between all known materials. Wax particles, for example, will agglomerate if dispersed in toluene at room temperature without any dispersant, and wax coated alumina particles will also agglomerate.

---

45 W. R. Cannon and R. Becker, Advances in Ceram., 26 (1989) 525.

If menhaden oil, which is a liquid and therefore is "softer" than wax, is put into the system, then a fairly stable suspension of wax (or wax coated alumina) can be made without agglomerates. Those oil-coated particles that do make contact will break apart easily from thermal Brownian motion, even if oil-to-oil van der Waals bonds temporarily form between the liquid coatings, because the oil coating is soft. It is probably not an accident that all the effective dispersants which operate by steric hindrance are either liquids or extremely soft solids like stearic acid. Therefore it appears that it is not sufficient to coat the particles with another material to diminish the attractions: the other material must either be a liquid or a very soft solid.

Regarding the second point in the list above, the minimum thickness that is effective should be considered. In some literature,[39] a distinction is made between *semisteric* stabilization and a true *steric* variant, and this has become a commonly used nomenclature. In the former, the coating is so thin that it only neutralizes most of the van der Waals attractions between the particles, while in the latter it is much thicker, corresponding to about m.w. 10,000, and it acts as a sort of bumper to fend off other particles. Theoretical calculations of the electric field attractions in low dielectric constant media have led to these ideas. However, from a practical standpoint it should be noted that many excellent dispersants have molecular weights in the range of only about 1,000, and they work as well as the larger molecules, if not better. It is the opinion of the author that there is no qualitative difference between the larger and smaller dispersant molecules that is worthy of a new name. There does not seem to be any convincing evidence that small molecules work by a really different mechanism than the larger ones, except for quantitative differences in bridging and tangling. All appear to work primarily as *soft, steric* "bumpers," regardless of any theoretical differences in the calculated attraction forces.

The organic solvent itself can act as a dispersant to some degree. Nonpolar solvents will be slightly adsorbed onto the powder by weak van der Waals bonding, but slightly polar solvents like alcohols are adsorbed much more strongly. In the complete slip system containing binder, etc., the solvent has been shown to compete with the binder for adsorption sites on the powder.[46] After ball milling, much of the naturally adsorbed water on the powder surface is probably displaced by adsorbed solvent, since the aggressive wiping action of the balls is bound to remove any top layer, and mass action would favor quick readsorption of the solvent, rather than the other things that are present at very low concentrations (binder, water, etc.). Moderately large alcohol molecules used as the solvent might be expected to operate as a dispersant more effectively than very small ones.

In the author's laboratory at Rutgers,[47] silicon carbide powder ($15 m^2/gm$) was found to be somewhat dispersible in pure isopropyl alcohol without any other "dispersant," but not in water or in toluene. When a higher m.w. dispersant was added in addition, the solids loadings could be made still higher. The dispersant plus toluene was not as good a system for this purpose as was the dispersant plus alcohol. The difference was slight: more porous agglomerates were observed in the green body using toluene than alcohol, with a lower green density. A still larger solvent molecule, n-butanol, was found to be a more effective solvent than toluene for achieving a high solids loading with difficult-to-disperse powders such as β''-alumina.[48] Even higher solids loading (70 volume %) of half micron powder has been claimed by other workers using *tert*-butanol as the solvent.[49] Evidently the use of a large, fairly polar molecule as the solvent can be very advantageous.

As mentioned in the adsorption discussion of Chapter 6, making the dispersant only slightly soluble instead of very soluble in the solvent system tends to encourage more adsorption of the dispersant onto the powder. In the author's previous laboratory at AT&T, it was experimentally determined[44] that menhaden fish oil is a more effective dispersant for half-micron alumina when the solubility was only

---

[46] K. E. Howard, et al., J. Am. Ceram. Soc., 73 (1990) 2543.

[47] X. Chen, D. J. Shanefield, and D. E. Niesz, Am.Ceram. Soc. Bul., 69 (1990) 496; X. Chen, "Pressureless Sintering with Yttria and Alumina Additives," M.S. Thesis, Rutgers University (1991).

[48] A. Stanzeski, D. W. Scott, and D. J. Shanefield, Am. Ceram. Bul., 72 (1993) 218.

[49] M. J. Edirisinghe, et al., Ceram. Trans. 26 (1992) 165.

moderate than when it was completely soluble, from the experiments summarized as follows:

Solvent	Menhaden Oil Solubility	Viscosity of Ball Milled Slip
Trichloroethylene	Soluble	Excessive
TCE plus Ethanol	Moderately Soluble	Low
Ethanol	Insoluble	Excessive

In the case of the ethanol, the menhaden oil was not soluble enough to become distributed around the powder particles, and instead it segregated onto the inner wall of the ball mill. Since then, the author has on numerous occasions improved dispersion by adding a small amount of a second solvent in which the dispersant was not soluble. An example is with sodium hexametaphosphate in water, adding a very small amount of acetone makes the salt less soluble, and more of the salt then becomes adsorbed.

More menhaden oil must be present in solution than is actually adsorbed.[44] This is done so that mass action can drive the oil onto the powder surface, thus getting the maximum possible coverage in competition with the solvent, water impurity, etc. It is often true that two or three pph of dispersant must be used for optimum results, even though only one pph or less is actually adsorbed. That can also be true in aqueous systems, and it is an important point. Several studies of adsorption isotherms for menhaden oil and glyceryl trioleate have been reported, and infrared spectroscopy shows evidence of some chemical bonding to alumina surfaces when conditions are optimized.[50]

---

[50] E. S. Tormey, L. M. Robinson, W. R.Cannon, A. Bleier, and H. K. Bowen, in "Adsorption of Dispersants from Non-Aqueous Solutions," J. Pask and A. Evans, eds., Plenum Press, New York (1981) 121; P. D. Calvert, et al., Am. Ceram. Soc. Bul., 65 (1986) 669; R. J. Higgins, Ph.D. Thesis, MIT (1990) 138 [only available directly from MIT, not from the usual UMI microfilm service].

Other factors in predicting the effectiveness of a nonaqueous dispersant are analogous to aqueous dispersion. Somewhat polar groups are desirable on the dispersant molecule to make it become adsorbed, and usually they are Lewis acidic or basic groups, as discussed in Section 2.2. However, they can not be so polar as to interfere with solubility in the nonaqueous solvent, and salts usually would be too polar. (It should be noted, though, that some ammonium salts do dissolve in alcohols, especially methanol, since it is the most similar to water of all the alcohols.) After adsorption, the m.w. must be in the right range, so that the unbonded chain will stick out into the solvent far enough to fend off other particles, but not far enough to bridge or tangle (see previous section on aqueous dispersion). Also as discussed in the previous section, a variety of different groups on the molecule, and possibly a variety of different molecule sizes, will each provide more opportunities to find a favorable site on the powder surface. Theoretically this should be true for steric reasons such as the variously sized steps and etch pits on crystal faces (Fig. 6.3), and also because of the variously acidic and basic sites that exist on some ceramic powders (reference 20).

A detailed example regarding menhaden fish oil is instructive. This dispersant, originally discovered in the paint industry and borrowed by ceramists about 20 years ago,[51] is still used for ceramic processing by IBM, Coors, and other companies, because of its versatility and its ability to continue* working regardless of lot to lot variations in the powder, humidity changes, etc. Probably the durability of this process is due to the fact that the fish oil contains compounds of many molecular weights and functional groups, and if one constituent can not

---

[51] H. W. Stetson and W. J. Gyurk, U.S. Pat. 3,698,923 (1972) and 3,780,150 (1973).

* In spite of the above statement, which is only true relative to most other dispersants, the author could use up the whole space allowed for this section by telling "war stories," or more properly "fish stories" when related to menhaden oil, about little disasters he has seen or heard of, especially via patent licensing and consulting. The technology of organic additives for ceramics is of biological complexity, and there are many possibilities for things to go wrong and lower the yield to below 50 percent. Most of the problems eventually had solutions, so the worst nightmares are best forgotten and subconsciously repressed, except those that had a humorous component. Some of the latter are being "preserved in amber" by appearing in the footnotes of this book.

solve a problem, the other one will. Similar dispersants that are pure, single materials such as glyceryl trioleate appear to require more adjustment in amount of dispersant, milling time, etc. as the powder changes from lot to lot.

Menhaden fish oil is ordinarily used for dispersing ceramics, paints, and inks after refining and air treating operations done by the supplier. Analysis of a typical batch is summarized in reference 5, Table 30.4, and evidently it consists mainly of the esters of glycerin and unsaturated organic acids. While glyceryl trioleate is not a major constituent, glyceryl trilinolenate and other similar esters are present. (Linolenic acid is a C18 carboxylic acid that has three double bonds instead of just the one in oleic acid.) In some samples that have been analyzed,[52] the esters of shorter and longer chain acids than C18 were also present, as well as polymerized species and some saturated oils.

In a series of experiments which are described further in reference 5, several batches of Alcoa A-16SG alumina were ball milled for 24 hours in TCE-ethanol at about 30 volume %, using 1.7 parts per hundred of menhaden oil as the dispersant. The viscosity was low enough for tape casting, and the slip could be poured readily by gravity. When synthetic glyceryl trioleate was substituted instead of the natural oil, the results were usually good, but some batches showed thixotropic behavior with high viscosity at low shear rate, although it could still be used for tape casting. The results of this and the next series of comparisons are summarized in Table 8.2.

It can be seen from the table that corn oil is similar to the menhaden fish oil in its ability to deflocculate the fine powder. Also, its chemical composition is similar, although not identical (see reference 52 or Table 30.4 of reference 5). At any rate, the main functional groups and the general molecular weight ranges are similar. Food grade corn oil obtained in grocery stores, and also herring fish oil (not shown in the Table) have been used by patent licensees of AT&T, as well as by the author's consulting clients over the years, and the results have been almost as good as with menhaden oil, although not quite as free of variations. Similarly, synthetic glyceryl trioleate does work as a substitute for the natural oils, but the viscosity has not been as low at

---

[52](a) D. R. Lide, Handbook of Chem. & Phys., CRC Press, Boca Raton, FL (1993) 7-29; (b) W. Sigmund, A. Roosen, et al., Adv'd.Mat'ls., 4 (1992) 73

Table 8.2 Experimental Dispersants

Dispersant, 1.7 pph	Functional Groups	Molecular Weight	Resulting Viscosity
Menhaden Oil	C=C, COO	$\approx$1,000	Low
Corn Oil	C=C, COO	$\approx$1,000	Low, Thixo.
Glyceryl Trioleate	C=C, COO	885	Low, Thixo.
Glyceryl Tristearate	COO	891	Solid
Glyceryl Monooleate	C=C, COO, OH	357	Medium
Ethyl Oleate	C=C, COO	310	Solid
Oleic Acid	C=C, COOH	282	Solid
Glycerin	OH	92	Solid
Octadiene	C=C	110	Solid
Plasticizers	COO, OH	$\approx$300	Solid
Polyvinyl Butyral	-O-, (also residual COO, OH)	$\approx$30,000	Medium

Special Note:
In reference 5, at the bottom of Table 30.3, it is noted that 3.4 pph of menhaden oil (twice as much as usual) yielded excessive viscosity, possibly because of tangling (see pages 232 and 253).

low shear rates. Apparently the mixture of complex materials in the menhaden oil is closer to what is needed.

Evidently glyceryl stearate does not have the necessary *double bonds*, since it has about the same m.w. as the trioleate and the same ester linkages but does not work. Probably the π electrons in the trioleate make coordinate bonds to the alumina, possibly acting as Lewis bases with the aluminum being a Lewis acid. (The author does not have any direct evidence for this last point.[††])

Glyceryl monooleate evidently has the required functional groups, C=C and COO, but it does not have sufficiently *high molecular weight* for optimal results, although it does work to some extent. Ethyl oleate has an even lower m.w. and does not work at all, so the size appears to

---

†† Some evidence was, however, obtained for coordinate chemical bonds between the π electrons of polymeric C=C adhering to a metallic Lewis acid; see D.J. Shanefield and F. W. Verdi, U.S. Patents 4,351,697 (1982) and 4,582,564 (1986).

be quite important in this particular series. The OH groups in the monooleate do not seem to be perturbing the size sequence from low viscosity to medium to solid.

Oleic acid has the π electrons and about the same m.w. as glyceryl monooleate but does not have the exact *ester* group that is needed, since it does not work as well as the glyceryl monooleate. Octadiene has two double bonds per molecule, but that can not overcome the disadvantage of lacking the other items.

It appears that, in this particular series of materials, three things are necessary:

(a) Unsaturation

(b) M.w. approximating 1,000.

(c) Ester linkage

It should be noted that these characteristics are not needed in all dispersants, since a number of completely different molecules work quite well with this powder and solvent mixture. Examples are the polyethyleneimine esters[53] with no double bonds, and higher molecular weights, etc. (See also polyethyleneimine discussions on pages 13, 231, and 287.) It is only in the series of similar materials investigated here that this combination is necessary. It should also be noted that polyvinyl butyral, ordinarily used as a binder, is a fairly good dispersant in this system, in spite of its much high molecular weight and quite different structure. Typically about 20% of it is really PVA (polyvinyl *alcohol*), and about 2% is polyvinyl *acetate*, left over from incomplete reactions in the polymer manufacturing process.[†] These

---

[53] For example, excellent nonaqueous dispersants are KD-2, -3, and -4 from ICI Co., Wilmington, DE. See J. F. Stansfield, U.S. Patent 3,996,059 (1976) [m.w. of KD-2 is approx. 1700]; A. Topham, U.S. Patent 4,224,212 (1980); E. DeLiso, and A. Bleier, in "Interfacial Phenom. Biotech. Mat'ls. Processing," Y. Attia, et al., eds., (1988) Elsevier, Amsterdam [analysis of KD-3]; L. Bergstrom, et al., Proc. 11th Riso Sympos. Metlrgy. and Mats. Sci. (1990) 193, published by Riso Natn'l. Lab., Roskilde, Denmark [KD-3]; E. Liden, et al., J. European Ceram. Soc., 7 (1991) 361 [KD-4 in cyclohexane].

[†] To manufacture PVB, acetylene (which is C_2H_2 with triple bonded carbons) is usually reacted with acetic acid to make vinyl acetate. This is then hydrolyzed with water to make vinyl alcohol, which immediately polymerizes. Butyraldehyde

polar groups probably aid in the adsorption of the PVB to hydroxyl-covered ceramic powders. In fact, they are probably also necessary for effective adhesion of PVB during its more common use as a binder: when comparing the special grades of PVB that have little residual polar "impurities," it turns out that they do not make as strong green bodies as ordinary PVB does. In most formulations it is observed that the addition of PVB binder lowers the viscosity, in spite of the fact that its high molecular weight might be expected to raise it, and PVB probably acts strongly as an auxiliary dispersant. Other investigators have also reported that PVB can be used as the only dispersant present.[54]   Also, ester groups on the binder molecule have been reported to be associated with lower viscosity slips, as if these groups might be causing the binder to be an auxiliary dispersant.[55]

Of the three characteristics listed above which appear to be necessary in some of the natural oils, the unsaturation is one which can easily be destroyed by excessive oxidation. Some steel drums of menhaden oil have been observed to lose their effectiveness after long periods of storage because of oxidation from air which leaked into the containers. A convenient quality control measure of the unsaturation is the "iodine number" test.[5] A weighed quantity of elemental iodine is dissolved in alcohol and slowly titrated into 100 gm of fish oil which has been dissolved in acetone. The first part of the iodine to be added reacts with the double bonds, as was illustrated in Fig. 3.15, and the reddish brown color of this iodine disappears. As the titration continues, the iodine color becomes visible immediately after all of the unsaturation is used up, and it remains visible. The weight of iodine added, in grams, up to this point of sudden color appearance is the "iodine number." Typically it is about 100 for air treated menhaden oil, but any significantly lower number would usually be associated with excessive oxidation, either by the supplier or during storage. On the other hand, a test for the degree of esterification (the "saponification number," which is similar but uses alkali addition and a

---

reacts with this to yield PVB. See W. L. Faith, et. al., "Industrial Chemicals," J. Wiley, New York (1965) 800.

[54] M. D. Sacks, et al., Advances in Ceram., 19 (1986) 175 [0.7% Monsanto B-79] ; V. L. Richards, J. Am. Ceram. Soc., 72 (1989) 325.

[55] Samukawa, N., et al., Gosei Jushi 36 (1990) 48 (in Japanese) [see Chem. Abstr. 113 (1990) item 28037t).

pH meter to determine the end point), or a test for the m.w. (gel permeation chromatography) only very rarely show any deviation from normal values.

Menhaden oil can contain up to 10% of some much more polar compounds such as lecithin, a phospholipid which is a complex fat containing phosphorous.[56] Lecithin could possibly ionize if there is a lot of alcohol present, and purified lecithin is used as an emulsifier in food products such as mayonnaise, so it might have significant effects on ceramic powder dispersion. It is difficult to analyze for lecithin, and therefore little attention has been paid to it in ceramic processing. A practical criterion for whether or not lecithin should be looked for in an oil is the degree of electrical conductivity of the solvent system being used. If it is essentially nonconductive, ionization is unlikely to be an important factor, unless only very small amounts of powder are being used, in which case the system might then be affected by small amounts of ionization.

Because the solvent and dispersant (and binder, later) are competing for adsorption sites on the powder surface,[24] large amounts of water in what are supposed to be nonaqueous systems can have strong effects.[57] (For comparison, it should be noted that very small amounts of water, chemisorbed on the surface in the form of OH groups, are usually not harmful and in fact are sometimes useful for promoting strong adsorptions of the dispersants, as in the cases of references 61 and 62 to be discussed later.) Cannon and colleagues at Rutgers[58] have reported the extensive microanalysis of trace water and its effects on nonaqueous ceramic slips. In nonaqueous systems of all kinds, the presence of several percent of water as an impurity often causes problems.‡ In the AT&T slip system described by reference 5, a water

---

[56] R. E. Johnson, et al., Advances in Ceram., 21 (1987) 323 [see pages 343,346].

[57] F. J. Micale, et al., Disc. Faraday Soc., Number 42 (1966) 238.

[58] W. R. Cannon, et al., Advances in Ceram., 19 (1986) 161; W. R. Cannon, et al., J. Am. Ceram. Soc., 73 (1990) 1312.

‡ One fine April day, the author and his capable technician were installing a nonaqueous process in the southeastern U.S. The process worked during this first day, but then while driving to the plant the next morning the technician noticed myriads of beautiful forsythia bushes just coming into bloom. The process utterly failed to work for the rest of that day. Soon after arriving home in disgrace at the northeastern U.S. research lab, the process did work as it always had, but the next day after that the forsythia blooms came out, as the summer weather crept

a water content of more than about 1% caused poor deflocculation. The solution to this problem in humid weather has been to do one or more of the following:

Bake the ceramic powder in an oven at 130°C overnight before use. Put it in the ball mill while still warm (about 30°C). Immediately add the rest of the materials and close the mill. This baking treatment will not remove all the adsorbed water, but it does eliminate what physical chemists call the "labile" water, that is, the adsorbate that is easily removed. These top layers would have been scraped off during the later ball milling step and would have entered the solvent system.

Store the binder in a metal cabinet (such as a solvent storage safe ty cabinet, which is suitably airtight), with nitrogen or dried air constantly trickling through it, for at least one week before use. By a mass action effect, this removes the approximately 1% water vapor which can permeate the "free volume" of most plastics (see bottom of Table 8.1), even ones that are not wetted by liquid water.

Use only alcohol that is anhydrous, and do not leave the containers open for long periods exposed to ambient air. Ethanol usually contains about 5% water, because that is an azeotrope which is difficult to eliminate, unless the material is specifically labeled "anhydrous."

For a long time it has been known by colloid scientists that dispersants in nonaqueous systems can sometimes react strongly with the powder surface to make true chemical bonds, not just enhanced van der Waals bonds.[59] Not only do small or medium size alcohol solvents react with surfaces, but also large alcohol molecules such as C18 which

---

inexorably northward. The process utterly failed. It turned out, after much agonizing, that a third factor had caused both the forsythia blooms and the process crash: it was a sharp rise in the humidity from nighttime rains. Draconian measures to remove moisture solved the problem in the lab and later in the factory, and the process was run successfully for many years afterwards.

[59] A. K. van Helden, et al., J. Colloid and Interface Sci., 81 (1981) 354; S Emmett, et al., Colloids and Surfaces, 42 (1989) 139.

are long enough to be thought of as dispersants can react, making stronger bonds and thus having greater effectiveness.[60]

Even more durable chemical bonds[61] can be made by reacting the powder with silane derivatives, which are partly inorganic and partly organic. These are quite effective in producing ceramic slips at fairly high solids loadings.[62] Some silica will be left in the fired body, but usually only a fraction of a percent. The silane derivatives are compounds of a general type which will spontaneously react with a hydroxyl-covered ceramic powder in a reaction written as follows, where -MOH represents the powder surface and the $(C_2H_5O)SiR_3$ might be something like ethoxytrimethylsilicon:

$$-MOH + (C_2H_5O)SiR_3 = -MOSiR_3 + C_2H_5OH.$$

The ethanol vaporizes away, and the silane compound is now covalently bonded to the surface through an ether oxygen linkage. The R groups could be long alkyl chains which are very compatible with organic solvents, and there is no longer as much chance for hydrogen bonding. Often these "silane treatments" are done with three ethoxy linkages to a single silicon atom, using only one R group. There are many varieties of such compounds,[63] sometimes referred to as "organometallic coupling agents." These are routinely used to enhance the adhesion of epoxy to glass fibers, for fiber reinforced plastic structures. Instead of a silicon central atom, a titanium or chromium atom or many other elements[64,65] can be used to modify a surface and make it water repellant but oil wettable. If the R is polar, it can be made to be water wettable.

Sometimes *silane treatment* is followed by the addition of an organic dispersant, and the two together result in higher solids loading than either alone. An example of this points up the extremely competitive nature of the modern technological world, and the need for

---

[60] F. F. Lange, et al., J. Am. Ceram. Soc., 77 (1994) 922 [UBE E-10 silicon nitride, refluxed in C18 alcohol at 200°C for 2 hrs.].

[61] H. Ishida , et al., Polymer Engrng. and Sci., 18 (1978) 128.

[62] K. Lindqvist and E. Carlstrom, J. Am. Ceram. Soc., 72 (1989) 99.

[63] E. P. Plueddemann, "Silane Coupling Agents," Plenum Press, New York (1991).

[64] A. Kerkar, et al., J. Am. Ceram. Soc., 73 (1990) 2879 (particularly page 2880).

[65] E. Liden, Ph.D. Thesis, Chalmers University of Technology, Goteborg, Sweden (1994) 49, 51.

a flexible attitude in manufacturing. In the 1970s, the author helped popularize the use of menhaden fish oil in making doctor bladed alumina substrates, as described in reference 44 and elsewhere. These substrates were then laminated together to make very successful ceramic "packages" for electronic components. However, polymeric ("plastic") packages began to compete with ceramics in the 1980s, because of lower cost, and it became obvious that trying to hold back the growth of plastic packaging, in order to protect the competitive position of ceramics, was like building a sand castle to hold back an incoming ocean tide. The author was then asked to help AT&T's plastic supplier to lower the thermal expansion coefficient of epoxy packaging material, in order to prevent "wire bond" breakage during temperature cycling. The solution to the problem was to obtain a very high solids loading of fused silica powder (low thermal expansion coefficient) in the epoxy. The means of doing this was to first silane treat the silica, and then adsorb menhaden fish oil onto it.[66a] This method was then widely copied, with various proprietary improvements, by other suppliers of epoxy to the electronic packagers,[66b] and the resulting plastic packages overwhelmed ceramics in the electronic packaging market. AT&T remained competitive by mostly switching over to plastics.[†] A happy ending for ceramists is that in the long run, in the 1990s, ceramics have been regaining ground because of the newer, very delicate and high value silicon "chips," and menhaden oil is still used to make the latest ceramic packages. (The author is now back in the ceramics field, but most assuredly not for that reason alone.)

An excellent type of partly inorganic dispersant is the phosphate ester family of compounds, which have several times been found to

---

66(a) Mark Bonneau (Dexter-Hysol Co.) and D. J. Shanefield (AT&T), unpublished work; (b) L. T. Manzione, "Plastic Packing of Microelectronic Devices," Van Nostrand Reinhold, New York (1990) [see particularly pages 87 and 89].

† The author was somewhat fearful when attending Ceramic Society meetings after that, although it turned out that these facts were not well known among ceramists, after all.

provide high solids loadings with otherwise difficult powders.[67] Some ceramists are concerned that phosphorus oxide will remain after firing, but this has not been actually reported to be a problem, since the esters are fairly volatile. It sometimes is found, however, that with the nonoxide advanced ceramics that can not be fired in air, those dispersants that work best do so because they are strongly bonded to the powder, and this prevents complete removal of all the organic material in nitrogen or argon. One example of this dilemma is that imidazoline, shown in Fig. 8.8, is one of the very few dispersants capable of working with ultrafine silicon nitride powder,[68] but it is difficult to remove. It is used commercially to prevent the carbon smoke residue of incomplete combustion in an automobile engine from agglomerating in the motor oil, being a popular additive for that purpose. However, it does not evaporate off from the ultrafine ceramic powder if it has to be fired in nitrogen. A potential solution to that dilemma is to fire in dry ammonia gas, which actually reduces the carbon to methane or other gases, instead of oxidizing it.[69] More will be discussed about ammonia "burnout" in the next chapter.

$$HO-(CH_2)_2- \overset{\displaystyle \overset{C}{/\ \backslash}}{\underset{\displaystyle \underset{C-C}{|\ \ |}}{N\ \ \ N}}-(CH_2)_7-\overset{\displaystyle \overset{H}{|}}{C}=\overset{\displaystyle \overset{H}{|}}{C}-(CH_2)_7-CH_3$$

8.8 Imidazoline, an unusually powerful dispersant, which is difficult to remove when firing in inert atmospheres. (As usual with organic structures, some of the hydrogens are not shown.) This structure was not particularly designed by chemists to look like this, but it happens to form from a condensation reaction between two simpler imines. It is available commercially as Unamine O®, from Lonza, Inc., Fair Lawn, NJ.

---

[67] P. Boch, et al., Am. Ceram. Soc. Bul., 66 (1987) 1653; W. R. Cannon, et al., Advances in Ceram., 26 (1989) 525; H. Yan, W. R. Cannon, and D. J. Shanefield, J. Am. Ceram. Soc., 76 (1993) 166 [with aluminum nitride powder].
[68] K. J. Nilsen, R. E. Riman, and S. C. Danforth, Ceram. Trans., 1A (1988) 469.
[69] F. K. van Dijen, et al., J.Eur.Cer.Soc., 5, (1989) 385; K. Drury, M.S. Thesis, Rutgers Univ. (1992).

Another dispersant borrowed for use in ceramics[70] from motor oil additive technology is Oloa 1200®, a product of Chevron Oil Co., which is polyisobutene succinimide. Succinic acid consists of two carbons with a COOH attached to each end, and the formal name is butanedioic acid. It can form an imide with a single nitrogen attached to the carbonyls, as shown in Fig. 3.13.

Although menhaden fish oil is an excellent general purpose dispersant for nonaqueous systems, and it is almost all organic, it is fairly difficult to burn out, possibly because of the strong bonding of the phospholipid constituents. The synthetic dispersants KD-2 and KD-3 (see reference 53) are easier to remove with heat, expecially if oxygen is not available for true "burnout," as in the case of nonoxide ceramics that must be fired in inert gas atmospheres. For use with semipolar solvents such as isopropanol, KD-2 was found to evaporate out easily in argon, although 800°C was required for complete removal.[71])

High Solids Loadings. Among the very high solids loadings that have been reported for nonaqueous slips recently is 60 volume % for A-16SG, but 74 volume % was achieved when some coarser powder was added to the composition.[72a] (The sinterability with the coarser powder added was not reported.) The solvent was hydrocarbon oil, similar to a very pure sort of motor oil. The dispersant was a stearic acid molecule onto which a C10 side chain was attached by obtaining the compound 12-hydroxystearic acid[72b] and esterifying decanoic acid onto the side of it, at the C12 position. This was done simply by refluxing the two chemicals together, and water is split off. (Refluxing is boiling with a condenser to return the vapors to the flask.) Experimental C5 and smaller, or C18 and larger side chains did not provide as good dispersion, and backbone chains as small as C4 were

---

[70] R. J. Pugh, et al., Colloids and Surfaces, 7 (1983) 183; F. M. Fowkes, Advances in Ceram. 21 (1987) 411 [Fig. 3].

[71] Xin Chen, "Pressureless Sintering of SiC with Yttiria and Alumina Additives," M.S. Thesis, Rutgers University (1991).

[72(a)] I. Sushumna and E. Ruckenstein, J. Mat'ls. Res. 7 (1992) 2884; (b) L. Bergstrom, C. H. Schilling, and I. A. Aksay, J Am. Ceram. Soc., 75 (1992) 3305 [see particularly page 3307].

also not as good. (The first table in Chapter 10 shows a calculation of vol. % solids loading.)

A 70 volume % solids loading of A-16SG in *tert*-butanol was reported[49] using sunflower seed oil as an additional dispersant. This is a polyunsaturated oil containing glyceryl trilinolenate and similar compounds. (Linolenic acid is a C18 structure with three double bonds.) The viscosity was fairly high, suitable for injection molding or extrusion. Linolenic acid itself, without being esterified, has been used to obtain 60 volume % of A-16SG.[73]

Linolenic acid and also linoleic acid (a C18 structure with two double bonds) are present in linseed oil, which is used in traditional paint formulations as a dispersant. They also function in paint as "driers," which are what a ceramist would call binders, in other words, they provide strength after the solvent (turpentine* in traditional paints) has evaporated. They spontaneously polymerize slowly in air, as oxygen causes the double bonds to link together in complex ways. Oleic acid and menhaden oil also polymerize in air, but very slightly, never becoming solid and only exhibiting a higher viscosity.

Since it has been noted that butanol can be the solvent and also the dispersant, and it is somewhat better than propanol, it should occur to the reader that alcohols with even longer carbon-carbon chains might make good solvent/dispersant liquids. This in fact turns out to be true. At such high solids loadings that the dispersant is really a lubricant (see discussion in connection with reference 31), a slightly more viscous material apparently prevents particles from touching each other, and it does this better than a "thin," lower viscosity material would. From the available literature, it is not possible to be sure of which m.w. is exactly optimal, because various conditions in the experiments to be compared were not always the same. However, it can be stated confidently that a C18 alcohol, octadecanol (also called stearyl alcohol) is one of the best solvent/dispersants at solids loadings of about 60 volume %, for half-

---

[73] J. Rives, et al., Ceram. Ind. (May 1991) 62.

* Turpentine consists of a group of fast evaporating, low m.w. compounds obtained by distilling pine tree sap. In traditional paint technology, the fluidizing liquid is referred to as the "vehicle," rather than the solvent. A higher m.w. liquid that is distillable from the pine sap is the slowly evaporating terpineol shown in Fig. 7.1. These are all quite good solvents for polymers. While the turpentine is used in paint, terpineol is used in ink.

micron alumina and similar fine powders.[74]   Although this might be somewhat new to many ceramists, the cosmetic chemists[75] have been using stearyl alcohol for years as a fluidizer/dispersant for submicron powders such as talc, titania, red iron oxide, etc.

Putting a double bond in the molecule at the 9 position makes oleyl alcohol, also called 9-octadecenol. In the author's laboratory,[76] this has been used repeatedly to produce injection molding compounds with 62 volume % of 10 m²/gm silicon nitride powder (UBE E-10), which was slightly better than could be obtained with saturated stearyl alcohol, all other variables being held constant. It appears that the double bond offers a small advantage. It was also used as a regular dispersant (2 parts per hundred) with n-butanol as the solvent, at the lower viscosities needed for tape casting, and the powder was ß''-alumina.[48]

Since Shanefield and Mistler had found that increasing the m.w. to about 1,000 improved the dispersion (references 5 and 44), and Cesarano and Aksay had found that decreasing the m.w. toward 1,800 also improved the dispersion (reference 27), this points to roughly 1,000 as an optimum for many (though certainly not all) systems. It seems quite probable that a smaller m.w. does not provide enough protection against Brownian motion collisions of the particles leading to agglomeration. Similarly, it seems probable that *higher m.w.* causes more *bridging.*†   Another point is that an *excessive amount* of dispersant (more than about 3 parts per 100 of powder) causes *tangling.* Either bridging or tangling can lead to flocculation. These appear to be important considerations when designing an optimized system.

---

[74] T. Sasaki, et al., Japanese Patent 62,278,160 (1987) [see Chem. Abstr. 108 (1988) item 117602n]; R. Gustafsson, et al., Swedish Patent 459,075 (1989) [see Chem. Abstr. 111 (1989) item 179658f]; F. F. Lange, et al., J. Am. Ceram. Soc., 77 (1994) 922 [UBE E-10, 60 vol. %, pressure filtered]; E. Carlstrom, D. Chalasani, and D. J. Shanefield, to be published [UBE E-10, 61 vol. %, milled first in a fugitive solvent, used for injection molding].

[75] H. Goldschmiedt, "Practical Formulas," Arco Publishing, New York (1978) 99.

[76] D. Chalasani and D. J. Shanefield, to be published.

† Rohm and Haas Co., Philadelphia, PA, has reported a new, low m.w. polyacrylate dispersant. See D. W. Whitman, D. I Cumbers, and X. K. Wu, Amer. Ceram. Soc. Bul., 74 (1995) 76.

There certainly are other factors (such as pH effects, chain branching, etc.) that can modify and even overwhelm the molecular weight effect. For example, diammonium citrate, a highly branched structure with a m.w. of only 226, has been used to achieve 63 vol. % solids with submicron alumina[77] at a pH of 9.

When very high solids are available, a very simple shaping technique[77,78] can be used: "gel-casting." The dense slip is poured into a mold, and then it is flocculated by a timed chemical reaction. Because of the high solids, the water does not all have to be removed from the relatively strong floc, in order to have sufficient green strength for the mold to be opened. Shrinkage is minimal, and the body tends to keep its shape during firing ("net shape"), and this might therefore become a very important ceramic manufacturing process in the future.

Possibly the highest solids loadings that are achievable with submicron powders are in the range of 60 to 70 volume %, because of the coating thickness effect discussed in connection with reference 29. It is remarkable that such high packing factors can be obtained at all, in view of the fact that dry pressing with a lubricant[79] does not usually reach these levels.

One potential application for high solids loadings is in achieving faster drying rates, especially where water is the solvent. The newly available high volume % slips might allow aqueous tape casts to dry fast enough for manufacturing relatively thick *substrate* and multilayer *packages,* instead of having to use organic solvents for these important electronic applications.

Those twin demons of ceramics, warping and cracking, should generally be diminished in the future, as the new high performance dispersion techniques are used more often. In fact, in a long series of development projects, it has been the experience of the author that shrinkage variations and large-pore defects were also diminished when improved dispersion (better dispersants and longer ball mill or attritor times) had been used. Emphasis on this point might prevent a host of problems for new workers in ceramics engineering.

---

[77] T. Graule, F. H. Baader, and L. J. Gauckler, ChemTech, 25 (1995) 31.

[78] M. A. Janney, O. O. Omatete, et al., U.S. 5,145,908 (1992); see also Ceram. Eng. & Sci. Proc., 15 (1994) 493; see also ref., 17(b), page 514.

[79] An exception is the example of the new superlubricant reported on page 278 .

9. BINDERS

The main purpose of the binder is to provide green strength so the body will simply maintain its shape until it is sintered. (Some ceramists and powder metallurgists use the word binder to mean all of the organics present in the green body, including the plasticizer, lubricant, etc. In paint technology, the strength function is provided by a material which is usually called the "drier, which was originally linseed oil and is now often an acrylic polymer.") In many cases the green ceramic body has to be machined, inspected, stored, and placed onto setters for firing, often by workers who are not perfectly careful, so considerable strength would then have to be provided by the binder.

In some other situations where not much green body strength is needed, the solvent alone has enough surface tension to bind the powder particles together sufficiently just by keeping the body damp, and examples are wet clays in the manufacture of simple shapes such as roof tiles, terra cotta drain pipes, rectangular bricks, etc. The more surface area, the more strength, and some extra fine clay can be added for making these products. However, "wet strength" alone is too limited for most other applications. Also, as the body dries during the initial stages of firing, it will weaken, and gravity might be sufficient to deform it, particularly if there is vibration from kiln car motion or fans.

Ball clay contains up to about 1% of naturally occurring *humic acid,** which acts as both a dispersant and a binder, and about 10% ball clay is added to many traditional ceramic compositions to provide dry strength from this natural organic binder. Most modern processes

---

* The term *lignite* is often used by ceramists to mean the same thing as humic acid. This naturally occurring organic remainder of decayed leaves and wood is sometimes divided by soil chemists into humic acid and fulvic acid, depending on whether or not it is soluble in alkali. However, the distinctions have no importance to ceramic processing, and the strength due to the natural organic content is best determined by direct experiment with the actual lot of clay being used. More information about ball clay is available in Appendix I.

make use of additional synthetic or other binders to improve the strength further and lessen the lot to lot variations found among the natural clays.

There are many plastics and glues that could be added to the slip in order to increase the strength of the green body, and epoxy would be a good example if strength were the only requirement. The characteristics listed in Table 9.1 remind the reader that the most important selection criterion in most cases is that the binder must be easy to burn out. Epoxy tends to cross link further if it is heated without air, becoming a tar and eventually carbon. It actually is used in a very few injection molded ceramics, where the shape is such that air can readily get in to all parts of the body and thoroughly burn the epoxy out[*] before the pores close during sintering, but these uses are rare. Fortunately, many other polymers that burn out better are available, so the easy burnout materials are used far more often as binders for ceramics.

Table 9.1 Desirable Characteristics of Binders

---

    1. Easy Burn-out

    2. Strong Green Body

       (a) Adhesion to Powder

       (b) Cohesive Strength

    3. Solubility in Fluidizing Liquid

    4. Low Cost

---

[*] For tape cast thin sheets, where burnout is easy but strength is important, Rohm & Haas Co. suggests adding a small amount of Jeffamine D-400 crosslinking agent to their Duramax 1050 thermoplastic emulsion binder, which strengthens it, gels it faster (effectively "drying" it faster), and prevents binder migration.

Turning to the second item in the table, if the binder is exceptionally strong, less of it has to used, and thus less has to be burned out. If enough binder is used to surround each powder particle, there does not have to be any adhesion to the powder; for example, polyethylene wax does not adhere well, but it can be sufficiently strong for some applications if a lot of it is used. On the other hand, where only enough binder is used to glue together the tiny spot where two particles touch, the adhesion can be a critical factor in the strength. In most cases the situation is closer to the latter example, and good adhesion is of some importance because it allows less binder to be used.

Binders do not necessarily have to be soluble in the "solvent," because most of them can be emulsified, using the methods described in the earlier section on cationic surfactants. Acrylate binders such as PMMA (Fig. 3.16) can be made from a monomer which has been emulsified in water,[1a] and the resulting polymer is then already emulsified, which makes it readily available for use in aqueous systems. The strength is ordinarily greater for the soluble binders, but it can be quite satisfactory for the emulsion types. A disadvantage of the soluble types is that viscosity tends to be high; a disadvantage of the emulsion types is that scrap green body (for example, punched out "window frames" in tape casting) usually can not be recycled back into the ball mill, because once dried, it will not disperse well again.[1b] Some newer emulsions are claimed to be recyclable, however.

Cost is an obvious item for consideration in choosing a new binder. In some products where the other characteristics are not important, it is the deciding factor. The natural products like starch and wax are most often the cheapest, and they are used where strength, etc., are not important issues.

After giving appropriate weighting to each of the characteristics in the table, the ceramist chooses which binder to use. A few of the most common choices, but by no means all the available choices, are shown in Table 9.2. Many other and possibly better binders could be used for each application. Further discussion on the decision making factors follows.

---

[1](a) I. Piirma, "Emulsion Polymerization," Academic Press, New York (1981). (b) K. Nagata, J. Ceram. Soc. Japan, 101 (1993) 845.

Table 9.2 Common Binder Choices

Application	Binders	Advantages
Slip Casting	Starch	Inexpensive
	Sodium Lignosulfonate	Inexpensive
	Na Carboxymethyl Cellulose	Inexpensive
	Sodium Silicate	Strong
	Ammonium Polyacrylate	High Solids
Tape Casting	Polyvinyl Butyral*	Strong
	Methacrylate Sltn. (in MEK)*	Easy burnout
	Methacrylate Emulsion	Easy burnout
	Ammonium Polyacrylate	High Solids
Extrusion	Methyl Cellulose	Heat -Gelation
	Starch	Inexpensive
	Sodium Silicate	Strong
Injection Molding	Wax and Polyethylene*	Easy burnout
	Epoxy or Phenolic*	Very Strong
Dry Pressing	Polyvinyl Alcohol	Plasticizable
	Methacrylate Emulsion	Easy burnout
Screen Printing	Alginates	Inexpensive
	Gums	Inexpensive
	Ethyl Cellulose*	Pseudoplastic
	Polyvinyl Butyral*	Easy burnout
Glaze Coating	Gums	Inexpensive
	Na Carboxymethyl Cellulose	Inexpensive
	Sodium Silicate	Strong

* Nonaqueous. (The others are aqueous.)
Note: Sodium compounds leave residue after burnout, and sodium silicate leaves even more residue.

9.1 Burnout

General Considerations. Figure 9.1 shows the weight of a green body while the "binder" (really binder, plasticizer, residual solvent, and adsorbed/absorbed water) is being removed during the first part of firing, or during a simulated pre-firing experiment in a TGA apparatus. What remains is the powder, and if burnout is complete, the solid curve will come down to exactly that weight, which is shown by the dashed line. However, this is difficult to measure accurately in practice, because the powder itself loses adsorbed water vapor, or in the case of a nonoxide powder such as aluminum nitride, it can gain weight when some aluminum oxide is formed on the surface. Residual carbon ("char") is best measured by direct analysis of the burned-out ceramic, using[2] a Leco® or other technique.[†]

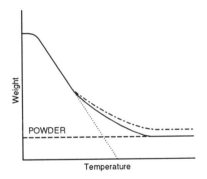

Fig. 9.1 Thermogravimetric analysis (TGA) of the binder plus the ceramic powder. Without the powder, the dotted line shows the same starting weight of binder alone dropping more rapidly. The dash-dot line shows a non-evaporating binder being fired in nitrogen

[2] H. Yan, W. R. Cannon, and D. J. Shanefield, J. Am. Ceram. Soc., 76 (1993) 166.

[†] In special situations a small amount of char is desirable or even necessary. Injection molded parts are sometimes made with a trace of ethyl cellulose added to the binder, in order to make enough char to hold the powder together during and after the major wax component is being removed. Extruded tubes are sometimes fired in a vertical position, hanging from the top ends, to keep them straight, and residual strength is needed for a while (from protein gelatin) after the major methyl cellulose binder has burnt out.

The plasticizer, being a liquid, can evaporate out at lower temperatures than would be required for actual burning. For this reason the first part of the curve shows mostly the loss of plasticizer and residual solvent, together with only a small loss of binder. The later part of the curve shows mostly binder loss and a small loss of residual plasticizer. These are merged and can not be resolved with a sharply defined boundary, even by taking the derivative of the curve, although a general distinction between the two regions can usually be discerned.

The same weight of binder, without the ceramic powder, loses weight along a different final path shown by the dotted line. However, when the powder is present, as illustrated by the solid line going up to the higher temperatures, the binder is usually adsorbed to some degree onto the powder surface, often with some chemical bonding, and therefore the last part of the binder only comes out slowly.

In air or pure oxygen, the last part of the curve represents burning, but if what we call "burnout" has to be done in nitrogen or argon, all of the binder removal has to be accomplished by evaporation. With wax, air burnouts look about the same as nitrogen "burnouts" on a TGA curve, because the wax evaporates equally well in either atmosphere. However, with cellulosic binders the nitrogen case looks like the dash-dot line in the figure, because the material does not completely burn out. Some binders like PVA will go all the way to zero weight in nitrogen when there is no ceramic powder present, but they switch to the dash-dot line behavior with powder, because then they are chemisorbed via H-bonding.

When a large amount of plasticizer is used in the composition, its evaporation leaves many pores for the air to get into and the carbon dioxide and water vapor combustion products to get out through. This can enhance burnout a great deal.[4] It is one of several reasons why a large amount of plasticizer should almost always be used with fine powders where there is a lot of adsorption. In one case, to the surprise of the author, TGA indicated that a particular plasticizer (butyl benzyl phthalate) actually impeded binder removal, possibly because of chemical reactions with the binder, but other plasticizers did not cause this problem.[3]

---

[3] H.Yan, W. R. Cannon, and D. J. Shanefield, Matls. Res. Soc. Proc., 249 (1992) 377.

Theoretical models for binder burnout have been proposed,[4] and many of the considerations for optimum removal have been reviewed in general.[5] There are special methods for improved burnout such as the use of a feedback controlled heating rate, with a constant weight loss rate instead of the usual constant temperature rise.[6] It is common practice for the temperature rise in ceramic and powder metallurgy firing operations to be halted at about 500°C for about one hour, and then the rise is resumed up to the sintering temperature. The heating "hold" or multiple[7] holds give good results in most cases, but crosslinking or other problems can occur because of deleterious hold sequences.[8] It is often found that a single binder with a narrow range of molecular weights will come out too suddenly, exerting excessive pressure from inside the green body and thus causing cracking. For this reason, a wide range of molecular weights, or even a mixture of different molecular species, is useful to make the outgassing occur over a wider temperature range. The Rohm and Haas Company offers poly(oxazolidinyl)ethyl methacrylate as a binder for ceramic and powder metallurgical use because of its unusually wide temperature range for unzipping/evaporation (150° to 600°C). While the ceramist ordinarily transmits the heat for binder burnout via conduction and radiation from the inside of the kiln, some advantage has been reported for the application of the heat via microwaves, so that it is uniform throughout the inside of the body instead of being too intense at the outside boundary and too weak inside.[9]

Burning. All organic materials can be burned out if there is sufficient air exposure and temperature. One problem arises when the green ceramic body is so large that both the oxygen diffusion inward to the binder and the outdiffusion of combustion products are slow. Another problem is where a nonoxidizing atmosphere is required in order to protect the ceramic or its metallization, and of course the binder can not burn without oxygen.

---

[4] M.J. Cima, et al., J. Am. Ceram. Soc., 73 (1990) 575, 2702.

[5] N. Grassie, "Chemistry of High Polymer Degradation," Interscience, New York (1956); H. M. Shaw, et al., Am. Ceram. Soc. Bul., 72 (1993) 94.

[6] E. Carlstrom, et al., Advances in Ceramics, 9 (1984) 241; E. Carlstrom, et al., U.S. Patent 4,534,936 (1985).

[7] J. A. Mangels, et at., Advances in Ceram., 9 (1984) 226.

[8] G. Scheiffele and M. D. Sacks, Ceram. Trans., 1A (1988) 538, 549, 559.

[9] E H. Moore, Ceram. Sci. and Engr. Proc., 13 (1992) 1081; X. D. Yu, et al., Mat'ls. Letters, 14 (1992) 245.

There are several potential solutions to these problems, other than using a binder that evaporates, which will be covered in the next section. For thick bodies where oxygen diffusion is slow, burnout in pressurized air has been reported to aid in binder removal.[10] This has to be done carefully, because excessively fast burnout can cause cracking or other defects, in any atmosphere.[11] Another solution is to add about 100 parts per million of a transition metal such as manganese or palladium, which can catalyze oxidation considerably.[12] The catalytic effect was initially discovered by accident when trace Mn was added to provide a dark purple color, and binder burnout problems were alleviated. The effect is so powerful, however, that it can cause the excessive oxidation rate mentioned above, if too much Mn catalyst is present.[13]

Another problem solver is to use a self-oxidizing binder such as cellulose nitrate, which is soluble in ethyl acetate or methanol, but unfortunately not in the more popular solvents. If some other binder must be used, an oxidizing salt such as ammonium nitrate or slightly acidified aluminum nitrate can be added to the slip. These are soluble in methanol or to a lesser degree in ethanol, and their solution can be poured into the ball mill just before starting to rotate it. If the main solvent causes slow precipitation of the nitrate, it will have been thoroughly mixed in before it can substantially separate. These nitrates decompose to give off oxygen at about 400°C, which is a high enough temperature to aid in burning organics but too low to destroy the metallization or aluminum nitride ceramic, etc. Nitrate oxidizers can be used with thick green bodies, or in nitrogen or argon atmospheres, or in vacuum firing.

Ceramics or powder metallurgy parts that need to be fired at high temperatures in hydrogen can make use of water vapor to oxidize the binder. So-called "wet hydrogen," saturated at a moderate temperature with water (having a "dew point" about 30°C), will have enough oxygen in equilibrium with the hydrogen and water at 1600°C for the carbonaceous material to burn off. In other words, water starts to decompose above about 1,300°C, but carbon monoxide does not, so the CO will form from the combustion, and the reaction will go forward. Some types of metallization such as tungsten will not be harmed by oxidation

---

[10] J. A. Mangels, Am. Ceram. Soc. Bul., 73 (1994) 37.

[11] G. Geiger, Am. Ceram. Soc. Bul., 72 (1993) 54 (see particularly page 60).

[12] J. Brownlow, et al., U.S. Patent 4,474,731 (1984); Y. Takeuchi, J. Hybrid Microelectr., 11 (1988) 93; S. Masia, et al., J. Mats. Sci., 24 (1989) 1907; J. Mats. Sci., 26 (1991) 2081.

[13] T. Negas, et al., Am. Ceram. Soc. Bul., 72 (1993) 80 [particularly page88].

under these conditions, because their oxides decompose just like water, as long as the oxygen concentration is low. Wet hydrogen is used extensively in manufacturing alumina insulators plus tungsten or molybdenum electrical conductor lines for IC packages.[14] Wet argon can also be used, if hydrogen presents a safety problem.[15]

Instead of oxidizing the carbon, it can be reduced to methane. However, hydrogen does not significantly react with carbon at ordinary sintering temperatures, probably because it does not adsorb readily to most materials. The dry ammonia gas molecule, NH_3, is large and polarizable and therefore does become adsorbed onto the binder, and it reacts to turn the binder into either methane plus nitrogen[16] or methane plus HCN gases,[17] depending on the conditions. The HCN can be trapped by bubbling it through NaOH solution, or it can be burned. The powerful dispersant for ultrafine powders, imidazoline (Fig. 8.8), has been "burned" (actually reduced, not oxidized) off of silicon nitride this way at Rutgers, as described in reference 16.

The products of combustion from binder burnout often include incompletely burned "smoke," which should not be vented directly into the environment unless the smokestack is far from any people who might be downwind. One method for completing the combustion all the way to safe carbon dioxide and water vapor is to add an auxiliary natural gas burner to the stack. This can be done by almost any engineer, unless there is an unusual problem, in which case Appendix III lists information sources for obtaining professional aid. (For example, see advertisements in Environmental Protection magazine). One must be careful in some cities to avoid excessive temperatures in the burner, however, so that NO_x is not produced inadvertently from the nitrogen of the air. Professional manufacturers of afterburners for smoke abatement are now installing expansion funnels to adiabatically cool the flame somewhat and thereby decrease the generation of NO_x in the flames. Recent environmental research indicates that NO_x is more likely to be producing atmospheric ozone when irradiated with

[14] H. W. Stetson, U.S. Patent 3,189,978 (1965); R. Tummala, et al., U.S. Patent 4,234,367 (1980); V. A. Greenhut and R.A. Haber, in "Ullmann's Encyc. of Indus. Chem.," VCH Publ., Deerfield Beach, FL(1985) vol. A6, page 55; D. R. Wall, et al., J. Am. Ceram. Soc., 73 (1990) 2944.
[15] M. J. Cima, et al., Ceram. Trans., 1A (1988) 567.
[16] K. Drury, M.S. Thesis, Rutgers Univ. (1992).
[17] F. K. van Dijen, et al., J.Eur.Cer.Soc., 5, (1989) 385.

ultraviolet light from the sun than are the volatile organic compounds (VOCs) which have previously been blamed.[18]

Another potential solution to the smoke problem is to catalytically enhance the oxidation in the flue pipe,[19] just as it is done in the catalytic converter of an automobile. This equipment would be difficult for a nonspecialist to install, but it is effective and is being used by chemical manufacturers.

Evaporation. Ceramics that must be protected from oxidation at the fairly high temperatures of binder burnout (about 400°C) can not, of course, be fired in air. Therefore, binders like wax that can evaporate are useful in these cases. While the shorter chain wax binders can evaporate completely, longer chain polyethylenes will usually decompose to some extent before vaporizing, and they can leave a small amount of tar or carbon residue.[20] Very low residues of solid material have been reported for the burnout of some binders that are not in common use, such as polypropylene carbonate (QPAC-40, from PAC Polymers, Greenville, DE),[2,21] polyisobutylene (Vistanex" MMO-80, from Exxon),[22] PEG of high enough molecular weight to be a solid (Carbowax" 20-M, from Union Carbide),[23] water soluble but evaporable agar polysaccharide (for efficient, low-organics gelling) in injection molding,[24] water soluble polyvinyl pyrrolidone,[25] and water-emulsified wax (X-5180, from Shamrock Technologies, Newark, NJ).

In searching for new binders with advantageous properties regarding easy burnout, one might start with start by reading the literature on some of the best materials in common use now. Appropriate examples might be the following: the chemistry of PVB,[26]

---

[18] R. J. Farrauto, et al, Chem. & Engrng. News, (Sept. 7, 1992) 34 [see p. 40].

[19] R. G. Silver, et al., "Catalytic Control of Air Pollution," Amer. Chem. Soc., Washington, DC (1992).

[20] Y. Tshuchiya, et al., J. Polymer Sci., Part A-1, Vol. 6 (1968) 415; D. W. van Krevelen, "Properties of Polymers," Elsevier, New York (1991) 459.

[21] J. Santangelo, U.S. Patent 4,622,240 (1986); M. Harley, Ceramic Industry (November 1995) 51.

[22] W. Farneth, et al., Matls. Res. Soc. Proc., 108 (1988) 95; L. E. Dolbert and J. D. Idol, Jr., U.S. Pat. 5,203,936 (1993).

[23] J. S. Reed, et al., Am. Ceram. Soc. Bul., 73 (1994) 61.

[24] A. J. Fanelli, C. P. Ballard, et al., U.S.Pats. 4,734,237 (1988) & 5,250,251 (1993); J.Am.Ceram.Soc., 72 (1989) 1833 [57v/o ZrO_2 used in making sensors].

[25] E.G. Nordstrom, et al., Am. Ceram. Soc. Bul., 69 (1990) 824.

[26] R.H. Fariss, ChemTech (September 1993) 38.

atactic polypropylene for injection molding,[27] and ethylene vinyl acetate copolymer (EVAC), also used commonly for injection molding.[28]

Several useful binders such as acrylates evaporate cleanly in nitrogen by "unzipping," that is, depolymerizing to oligomers that can evaporate. These are usually poly(methylmethacrylate) or variations of it. The vapors that come off have been analyzed by infrared and mass spectroscopies.[2,21,22,29] In spite of the fact that the pure binders vaporize, when they are adsorbed onto ceramic powder in a real green body the TGA curves during binder removal follow the dash-dot line of the figure.[30]

Other binders such as PVB do not completely unzip or evaporate when pure, and they do leave a small amount of residue even without the powder. They leave even more with the powder present.[2,31] Very small amounts of this solid residue can cause defects in the fired body.[32] In the subdivision of powder metallurgy which is called "thick film" metallization of ceramics for electronic applications,[33] the residue from incompletely burned out binder also causes defects, especially in the newer copper technology where nitrogen firing must be used.[34]

Because of the fine powders present in green ceramic bodies, even if the binders do evaporate, some types tend strongly to redeposit onto the powder. This might appear to be a primary residue, but it can be identified as a redeposition by running an experimental gas tube reactor with the separate steps of evaporation and redeposition kept several centimeters apart.[1,2,35]

A potential safety problem exists when the binder is evaporated, either completely as with wax, or partially as with PVB. According to an old saying, "What goes up must come down," and binder that evapo-

[27] T. Sasaki, et al., Japanese Patent 62,278,160 (1987) [see Chem. Abstr. 108 (1988) item 117602n.

[28] Ref. 27 also reports the use of ethylene-vinyl acetate copolymer (EVAC).

[29] R. J. Higgins, Ph.D. Thesis, MIT (1990).

[30] G. Scheiffele and M. D. Sacks, Ceram. Trans., 1A (1988) 538.

[31] G. Scheiffele and M. D. Sacks, Ceram. Trans., 1A (1988) 549, 559.

[32] M. J. Cima and J. A. Lewis, Ceram. Trans., 1A (1988) 567.

[33] D. J. Shanefield, "Electronic Thick Film Technology," page 284 in "Ceramic Films and Coatings," J. B. Wachtman and R. A. Haber, editors, Noyes Publications, Park Ridge, NJ (1993).

[34] D. Button, et al., Matls. Res. Soc. Proc., 72 (1986) 145.

[35] D. W. Whitman, Proc. Intl. Soc. for Hybrid Microelectr., (1988) 421;

D. R. Wall, et al., J. Am. Ceram. Soc., 73 (1990) 2944.

up inside the chimney, catches fire and melts, and droplets land on the roof.†

One solution to this problem is to install a trap to condense the oligomer vapors. This can be a simple, ambient air cooled, thin metal box. A metal baffle inside diverts the gas flow, and the oligomer will solidify on it. The box must be cleaned out periodically. Another solution, more expensive, is to run a water spray "scrubber" in the flue path. However, neither of these can be too far from the kiln, or the material might condense before it gets there.

Other Removal Methods. Injection molded ceramics and powder metallurgy green bodies often make use of wax as a binder, particularly for metals that would be damaged by oxidizing burnout. One of the popular ways to remove ("debinder") the wax is to barely melt it, and let it wick into a porous blotter that is in contact with the body.[36] This must be done very slowly, sometimes taking more than a week, so that the binder does not expand enough from overheating to separate the metal particles significantly. If they are separated, they will flow because of gravity, and the body will become distorted. Sometimes the blotting material consists of "grog," which is small pieces of porous ceramic, pressed against the body from all sides to support it and prevent "slumping." This can be done in a vacuum oven, to remove the rest of the wax by evaporation, or sometimes only vacuum evaporation is used.

Another method for removing binder, particularly with valuable products like injection molded rotors for automobile turbochargers, is to dissolve ("extract") the binder out of the green body using supercriti-

---

† The tape casting process described in reference 44 of Chapter 8 was licensed to many companies, and in the 1970s the author was working on solving various problems that the licencees might have. In the author's pilot plant facility, when the process was scaled up, large amounts of PVB vapor condensed and collected in the vent pipe and then caught fire. Melted, burning PVB drops went up the flue and landed on the tar-covered flat roof, which also caught fire. The author and his trusty technician climbed to the roof and put out the flames before the fire department arrived, but a few minutes later would have been too late. All the legally licensed companies were notified to trap the PVB residue and clean it out periodically. However, at least one company (in Texas, making use of illegal immigrant workers) was using the process without being licensed, was not notified, and had a disastrous fire from the same phenomenon. Nobody died, but those particular workers refused to return to that plant when it was rebuilt.

[36] I. Peltsman, et al., Interceram., 4 (1984) 56.

the wax by evaporation, or sometimes only vacuum evaporation is used.

Another method for removing binder, particularly with valuable products like injection molded rotors for automobile turbochargers, is to dissolve ("extract") the binder out of the green body using supercritical ammonia or carbon dioxide.[37] This is done at a high pressure, above the critical point of temperature and pressure, where there is no difference between the gas and the liquid phases: they are the same thing. Many materials such as carbon dioxide become unusually good solvents above their critical points and can be used to extract solid organics from porous ceramics, etc. Although this may seem like an expensive process, it has become cheap because of new pump technology, and it is now used to extract the caffeine from coffee, so that the caffeine can then be sold to cola drink manufacturers. The extracting fluid is carbon dioxide, which leaves no detectable residue, and which would be safe even if it were still there. This process replaces extraction of caffeine by methylene chloride liquid, which did leave a detectable residue (parts per billion) and which is a "suspected" carcinogen based on animal experiments. Supercritical extraction possibly will become more commonly used to remove binders in the future.

Nonuniform Burnout. If lumps of the binder fail to dissolve during processing, they will leave pores after burnout.[38] Lumps of binder or any other debris such as unbroken aggregates of powder can be filtered out of slip just before the shaping step, and it is highly recommended to do so. Well-milled slip can be pumped through very fine filters, and a simple design for making these filters has been reported,[39] in which screen printing sieves are clamped between two metal plates. Holes are drilled and tapped in each plate, pipe nipples are screwed into the holes, and plastic tubing is fitted over the nipples. Slip is pumped upwards through the filter, to avoid airlocking, as described further in the reference. Sieves can be obtained with 10 μm holes, from screen printing equipment suppliers.

In some applications it is desirable to leave very small lumps of high m.w. binder in the green body, to evaporate out or burn out during

---

[37] T. Sasaki, et al., Japanese Patent 62,278,160 (1987) [see Chem. Abstr. 108 (1988) item 117602n]; D. W. Matson, et al., J. Am. Ceram. Soc. (1989) 871; T. M. Sullivan, U.S. Patent 4,961,913 (1990).

[38] F. F. Lange, et al., J. Am. Ceram. Soc., 69 (1986) 66.

[39] D. J. Shanefield and R. E. Mistler, Am. Ceram. Soc. Bul.., 55 (1976) 213.

the first stages of sintering and leave a continuous network of pores. This is used to make ceramic filters.[40] In fact, these filters can be made with holes that are only of molecular sizes, by embedding insoluble PEG molecules in sol-gel alumina.[41] The PEG burns out during the last stages of sintering and leaves a network of ultrafine holes. Long holes can similarly be made in otherwise nonporous ceramic, to be used as buried pipes or channels for allowing cooling water or oil to be pumped inside a solid ceramic body. This is done by embedding narrow plastic rods in the green body before firing.[42]

If the binder is not evenly distributed in an ordinary green body, this can cause warping or cracking during firing, because the shrinkage will be greater where there is excessive binder.[43] One possible cause of nonuniform binder distribution is settling of the powder downwards and the consequent displacement flotation of the binder upwards. In tape casting, this is blamed for the fact that there is usually a little bit more binder at the top of the dried tape than at the bottom, which has been shown by TGA of a thin slice taken from the top having more than average weight loss (see reference 44 in chapter 8). Another possible cause of tape being "binder rich" at the top is "binder migration" where the solvent must move toward one surface in order to evaporate (in the case of tape cast material, it is moving toward the top), and it carries some dissolved binder with it. This binder is left behind when the solvent vaporizes.

In general, both types of maldistribution can be alleviated to some degree by using less binder, or early gelation of the binder.[44(a)] (Thermal gelation can be used.) Strong adsorption of the binder,[44(b)] fine powder, and high viscosity, all tend to prevent migration, but they usually can not be changed without sacrifices elsewhere.

Before leaving the subject of burnout, a word should be said about the possibility of using a "ceramic precursor" as the binder, where it purposely is designed not to entirely burn out but to make more ceramic in the pore areas instead. The binder would be an organometallic compound of silicon, nitrogen, and organic radicals, in order to leave behind silicon nitride, for example. While several of these have been

---

[40] L. C. Klein, et al., Ceram. Engrng. & Sci. Proc. 9 (1988) 1261.

[41] K. Maeda, et al., Chem. & Ind., 23 (1989) 807

[42] M. Kahn, et al., Advanced Ceram. Mat'ls., 2 (1987) 836.

[43] Y. Ting and M. J. Cima, Ceram. Trans., 26 (1992) 115.

[44(a)] G. Y. Onoda, Jr., in "Ceramic Processing Before Firing," G. Y. Onoda, et al., Eds., J. Wiley, New York (1978), 235 [see particularly page 248]. [(b)] Y. Zhang, et al., J. Am. Ceram. Soc., 79 (1996) 435.

reported in the literature for producing silicon nitride[45] and silicon carbide,[46] a large amount of "outgassing" (evolution of methane, etc.) has made it difficult to close the last pores, although progress is being made in this field of research. These organometallics are currently used to make fibers, where the surface area of the product is so high that the gases can escape without preserving the pores.

9.2 Adhesion

Three kinds of adhesion are shown diagramatically in Fig. 9.2. The van der Waals type is relatively weak and requires clean surfaces and a large area of contact. Chemical bonding, by comparison, is so much stronger that bonds can be made even if part of the area is not clean or flat. Mechanical interlocking is another possibility, if the surface is

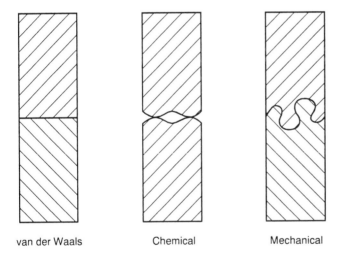

van der Waals   Chemical   Mechanical

Fig. 9.2 Three means of obtaining adhesion between the powder and the binder.

[45] Z. Zhang, et al., J. Am. Ceram. Soc., 74 (1991) 670.
[46] K. Shimada, et al., Japanese Pat. 63,206,354 (1988) [see Chem. Abstr. 110 (1989) item 62539f].

rough. Generally in the adhesion of binders to ceramic powder, H-bonds are desirable, but van der Waals[47] can also be strong enough if the surface has just been scraped clean by the powerful wiping action of sliding ball mill media.

PVB is well known in the polymer industry for its high adhesive strength to glass surfaces. The ether oxygens and also the residual PVA and acetate groups provide enough polarity to make strong H-bonds to the chemisorbed water on the glass.[26] It is used extensively to make automobile safety glass by gluing together pairs of window panes. In ceramics, it is used as a strong binder for alumina and other powders in tape casting, etc. Generally, good adhesives are fairly polar materials. In the section on adsorption in Chapter 6, it was pointed out that Lewis acid/base interactions and other factors can lead to good adsorption, and those same factors also lead to good adhesion.

In ceramic processing, the binder solution is ordinarily milled or attritted together with the well-dispersed powder for a long enough time for some of the adsorbed water vapor and solvent to become displaced by adsorbed binder. In reference 44 of chapter 8, it was reported that a certain minimum milling effort must be made in order for the binder to become attached. More detailed analysis of milling sequences has been reported by Cannon and colleagues at Rutgers.[48] There is a certain maximum milling effort allowable (usually milling time) when the binder is present, and exceeding this will degrade the strength. The reason for the maximum is that too much aggressive shear can break the long binder molecules.[49] Therefore, the best strategy is to mill first with the solvent and dispersant, but without adding the binder yet. After suitable deagglomeration, the binder and plasticizers are then added, and a short time of further milling is done to provide for proper adsorption of the binder, which probably has to dis-

---

[47] In the early days of the U.S. space program, it was demonstrated in ultrahigh vacuums similar to those of outer space that flat, clean surfaces can van der Waals bond together with considerable strength. However, a small amount of adsorbed soft material spoiled the bonds. For this reason, astronauts' gloves are covered with a thin layer of grease, so they can let go of tools, etc., after holding them. See Anon., "Cold Welding in Space," Special Publication 431, ASTM, Philadelphia (1967).

[48] A. Karas, T. Kumagai, and W. R. Cannon, Adv'd. Ceram. Mat'ls., 3 (1988) 374.

[49] W. W. Lau, et al., "Modern Size Exclusion Chromatography," J. Wiley (1979) 225.

place some adsorbed water, other solvent, or dispersant.‡ Other workers have also found optimum milling sequences very similar to this.[50]

9.3 Green Strength

The green body strength has been estimated theoretically.[51] It has been measured[48,52] for dog-bone shaped sheets of tape cast green bodies after various milling and binder addition sequences, as discussed briefly above. Also, the higher the packing factor, the higher the green strength, other things being equal.[53]

The effects of Griffith flaws on the strengths of dry pressed bodies has been studied.[54] Usually, the same kinds of defects that would weaken the green body will also weaken the body after it is fired,[55] so a great deal of effort is now being directed toward inspecting for flaws of all kinds in the green state.[56] This is called either nondestructive testing (NDT) or nondestructive evaluation (NDE). It is being done exper-

---

‡ Colloid chemists do a demonstration experiment to illustrate competing adsorptions as follows. A new glass beaker is taken out of the shipping box and wiped with a dry towel but not washed. Fairly concentrated gelatin solution is put into it and chilled in a refrigerator until it gels. It is then left on the lab bench overnight, and its water evaporates. A much smaller cylinder of shrunken, dry gelatin is now in the beaker, and it can be rattled around inside, making a "clink, clink, clink" sound. Microscopic examination shows that the high pitched sound is from small fragments of glass, "shards," that the strongly adhering gelatin has pulled off from the inside surface of the beaker, which strike the beaker when rattled. Glasses are weak in tension, and the gelatin adhered strongly enough via H-bonds so that the glass shards were broken away. If the experiment is repeated, but the new beaker is first washed with detergent, the rattling sound is "clunk, clunk, clunk," because the adsorbed detergent did not allow the gelatin to get a grip on the glass itself, and there are no microscopic glass shards on the dry gelatin cylinder. In tape casting, this powerful shrinking force during the drying of gelled PVB pulls the ceramic powder particles together with the same effectiveness as dry pressing, yielding a packing factor of about 55% (compare Tables 5.1 and 8.1), but the dispersant must be displaceable by the binder during a long enough second stage of milling.

50 A. Roosen, Ceram. Trans., 1B (1988) 675 (see particularly Fig. 10).

51 G. Y. Onoda, J. Am. Ceram. Soc., 59 (1976) 236.

52 S. Forte, J. R. Morris, and W. R. Cannon, Am. Ceram. Soc. Bul., 64 (1985) 724; T. Chartier and A. Bruneau, J. European Ceram. Soc., 12 (1993) 243.

53 K. Nagata, Ceramic Trans. , 22 (1991) 335 (see Fig.4).

54 K. Kendall, Powder Met., 31 (1988) 28.

55 D.J. Cotter, et al., Ceram. Engrng. & Sci. Proc. 9 (1988) 1503.

56 L. M. Sheppard, Am. Ceram. Soc. Bul., 70 (1991) 1265.

imentally, and to some extent in production, by ultrasound imaging,[57] and by x-ray examination of various kinds.[58] In an effort to find the smallest flaws, which are not visible by the more accessible inspection methods, very sophisticated techniques have been tried, examples being nuclear magnetic resonance (NMR) tomography (lately called magnetic resonance imaging or MRI)[59] and even neutron scattering.[60]

To some degree, the strength can be increased by using a binder of higher molecular weight (because of increased *tangling*), but so many factors of polarity, competition with adsorbed dispersant, etc. come into play that determining the strength is mostly a matter of trial and error. For example, PVB has an excellent reputation for reproducibly high strengths, but for very thin sheets of tape cast green ceramic, some emulsified polyacrylate binders have been just as strong, even though theoretically the emulsion should not be able to provide as much adhesion to the particle surfaces as the true solution of PVB does.

The amount of binder influences the green strength in an obvious way: the more binder, the more strength. Figure 9.3 shows the bulk green density as more binder is added. Since it fills in pores without raising the outer boundary volume (Fig. 8.3), the density is raised. On the other hand, the apparent density goes down, because the measured volume increases, and low density polymer is making up more of the total composite material. When the two densities merge, the pores are filled. That merging point is approximately the point of maximum strength, since the ceramic powder acts as a "filler" in the plastic, and this usually makes a stronger material than pure plastic.

In practice, things are not so simple. Experiments have shown that additional binder close to the merging point does not fill the last pores, but instead floats to the top of the body, at least when tape casting is done.[61] Therefore, to avoid nonuniform binder distribution and its attendant warping and cracking, experience leads to the use of slightly less binder than the merging point of the two densities, so that binder float is minimized. The strength of the tape is only degraded slightly by being a little bit "binder starved."

---

[57] M. C. Bhardwaj, Am. Ceram. Soc. Bul., 69 (1990) 1490.

[58] D. H. Phillips, et al., Am. Ceram. Soc. Bul., 72 (1993) 69; K. Isozaki, et al., Am. Ceram. Soc. Bul., 72 (1993) 95.

[59] L. Cartz, "Nondestructive Testing," ASM Int'l., Materials Park, OH (1995).

[60] B. D. Sawicka, et al., Ceram. Engrng. & Sci. Proc. 9 (1988) 1491.

[61] See chapter by Mistler, Shanefield, and Runk, page 411, in "Ceramic Processing Before Firing," G. Y. Onoda and L. L. Hench, eds., J. Wiley, New York (1978) [particularly Fig. 30.14].

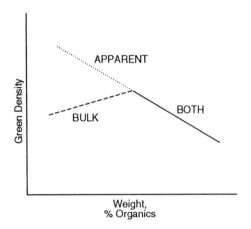

Fig. 9.3 Effects of increasing binder content on green densities.

Plasticizers. The plasticizer mainly toughens the binder. It is a liquid oligomer, structurally similar to that of the binder, and it therefore can penetrate in between the binder molecules and lubricate them, allowing them to slide against each other and take up strain without actually cracking.[†] Many different types used in ceramics have been listed by Williams.[62] Very often two plasticizers are much better than one, possibly getting into different size spaces between the binder molecules. Some binders are more readily plasticized than others, with an example being the fact that polyvinyl acetal is not generally plasticizable, while polyvinyl butyral is, which is why we use PVB.[25] PVA

[†] A polymer with a Tg below room temperature is equivalent to being self-plasticizing, and it often will not need the addition of another material to make it less brittle. An example of such a material is Rohm and Haas poly(oxazolidinyl)ethyl methacrylate.

[62] J. C. Williams, p.173 in F. F. Y. Wang, "Treatise on Materials Science and Technology, Vol. 9," Academic Press, N.Y. (1976).

is extraordinarily plasticizable by glycerin, which is why we use it for dry pressing where it must be distortable without cracking. Water is an excellent second plasticizer for PVA, and a large amount of it should be used with the glycerin (see next chapter example). Often it is advisable to use more weight of plasticizer than of binder. Without the ceramic powder, this would become a sticky, weak mass. However, the powder surface area has a "drying" action, and with the powder the whole composite is strong and not sticky. Without the plasticizer, and even with three times as much binder, the green body usually is too brittle, at least when fine powders are used.

Just as a small amount of liquid can hold powder particles together by capillary action, a larger amount can act as a fluidizer and let the particles move freely. Similarly, a small amount of plasticizer makes powder sticky by capillary action, but a larger amount can act as a "release agent" by fluidizing the interface between the particles and the mold or support film.‡ An intermediate amount can do both, depending on the exact situation. In one case, working with tape cast fine powder and using a PVB binder, stearic acid (2.8 parts per hundred) functioned as a powerful auxiliary plasticizer in addition to the usual liquid plasticizer, thus preventing cracks.[63] This was done with very small batches of slip, which tend to dry too fast and would therefore crack without the auxiliary effect. In addition to plasticizing, the stearic acid also acted as a *release agent* to allow the shrinking tape to slide instead of sticking to the supporting film and cracking, but it had just the right stickiness to prevent the drying tape from releasing too early and from rolling up into a cylinder. This is an example of the complex balance of functions that a plasticizer sometimes must provide. It can be seen from this example that finding the right balance of binder strength, release, and other properties by adjustment of one or two plasticizer amounts and types can be a difficult endeavor. A large block of development time should be budgeted in advance whenever

---

‡ A release agent aids in removing the freshly cast green body from the mold or support film without sticking. For examples of some release problems and their solutions, see Table 10.1. Although these are presented in the form of a "case study" involving tape casting, they are similar to the problems and solutions that occur in other casting and molding processes.

[63] D. J. Shanefield and R. A. Desai, U.S. Patent 5,002,710 (1991).

possible, in order to be sure of accomplishing these things in any new ceramic process.[*]

If a large amount of binder plus plasticizer is used, as is usually the case in tape casting, a valuable advantage is that several sheets of green material can "laminated," that is, pressed together (at approximately 1000 psi, 100°C, for 10 minutes), and they will stick together and fire to become one continuous body. Layers or just lines of powdered metal can be included between the ceramic layers, making "multilayer" composites.[64] (For simple formulations, see also Ref. 33.) These are used extensively in electronic capacitors, IC packages, etc., up to 100 layers thick. If difficulty is encountered in making the layers stick without delaminating during firing, sprinkling extra plasticizer on the mating surfaces before putting them together can alleviate the problem. In difficult situations, a mixture of solvent, binder, and plasticizer might have to be used. Sometimes this is done in vacuum, to prevent air entrapment.

In general, other advantages of using a large amount of liquid plasticizer include much easier binder burnout (as mentioned previously), in addition to much easier release from any sort of mold or supporting film. The Processing Examples chapter will illustrate some formulations.

9.4 Other Additives

Homogenizers are liquids which are compatible with two other materials in the composition and can therefore keep them from separating. An example is that diethyl ether will not dissolve in water, so the two are incompatible. However, ethyl alcohol is soluble in each of them, so a mixture of the three can be made in which they are all three mutually soluble. Cyclohexanone has been listed in some old ceramic formulation patents, and it is occasionally used in modern formulations[2,48] also. It tends to prevent skin formation during drying, since it keeps everything in solution longer than usual (see also page 181).

---

[*] In any ceramic development project, it is well to be reminded of Shanefield's Law, which reads, "Everything takes longer than you thought it would."

[64] H. W. Stetson, U.S. Patent 3,189,978 (1965); W. J. Gyurk, U.S. Patent 3,192,086 (1965); R. A. Gardner, et al., Solid State Tech., 17 (May 1974) 38; A. J. Blodgett, Sci. Amer. (July 1986) 86; H. W. Stetson, in "Ceramics and Civilization, Vol. III," Am. Ceram. Soc., Westerville, OH (1987), page 307.

Wetting agents are the same thing as surfactants, which is a broader category than dispersants. In some older literature, dispersants are referred to as wetting agents, as well as deflocculants, surfactants, etc.

Lubricants are important for uniform compaction in dry pressing and extrusion, where very high viscosity material is pushed hard against metal dies, and yet it must slide without sticking. The author recommends that they be used in those two processes, if for no other reason than to decrease die wear, but they often are not used. Other advantages are freedom from cracks in the green body due to stick-slip motion in the die and of warpage due to nonuniform compaction. A higher overall green density can almost always be achieved when good lubrication is used. As discussed briefly in the section on dry pressing in Chapter 10, both an "internal lubricant" (between the ceramic particles) and and "external lubricant" (between the particles and the stationary metal die) should both be used whenever possible.

After the shaping function has been accomplished, and then during the final "ejection" part of extrusion (or dry pressing and similar shaping processes), what is desirable is *slippage,* or *plug flow,* as illustrated at the bottom of Fig. 6.11. The *external lubricant* particularly helps in achieving this, with a minimum of shear-warping during that final part of the process. The ratio of slippage to shear can sometimes be estimated by the Benbow equation described in reference 23, although it might be difficult to make practical use of this.

Hydrocarbon oils and waxes are fairly effective lubricants for ceramic processing, but in some instances stearic acid appears to be slightly more effective. Wax is cheaper, however, and a water emulsifiable form of it can be obtained from Shamrock Technologies in Newark, NJ. Somewhat less stable emulsions of wax or other oily substance can be made by ceramists for their own use, by following the procedure on page 223 of Chapter 8 (also summarized in the dry pressing section of Chapter 10). Improved stability can then be obtained by passing the emulsion through a tabletop homogenizer[**] or emulsifier machine, which can be purchased from laboratory suppliers. These machines have pumps which force the emulsion through extremely small holes, making the particles of oily substance all become smaller than that size. Factory size machines of this type can also be obtained.

[**] Not related to the "homogenizer" chemicals described above.

Much research has been published about lubricants for engines, but most of it relates to resistance to burning in the cylinder, or contamination by gasoline, etc., which is not of great interest to ceramists. Differences in chemical structure have made only very slight difference in lubricating effectiveness.[65] Carbon chains of about C10 to C18 length appear to be among the best materials.[66] In one situation, the polyisobutylene (STP® additive) that is sometimes used to raise the viscosity of motor oil was found to be an excellent lubricant for dry pressed alumina.[67] Being a highly branched chain, it has a high viscosity, as was mentioned in the discussion of Fig. 3.19. In order to use it in an aqueous system, it was emulsified with a cationic "quat" surfactant. However, a complicating factor in assessing its effectiveness as a lubricant is the quat by itself is also a lubricant, and it was difficult to determine which chemical was doing most of the lubricating.

Aluminum stearate and ammonium stearate were slightly better than stearic acid as lubricants in ceramic dry pressing.[68,69] Marschner, in reference 69, reported that stearate ion can make a chemical bond to alumina, as determined by infrared spectroscopy. Salts of stearic acid, being much more ionized‡ than pure stearic acid itself, acted as slightly better lubricants in her experiments. She also found that when stearic acid was emulsified instead of being made into a salt in order to use it in aqueous systems, the quat emulsifying agent was itself a good lubricant, as were most of the usual dispersants. Therefore she ran several experimental series without either of those, using sodium metaphosphate instead of polyacrylate dispersant and salts of stearic acid instead of emulsions. This points out the need for special care which must be observed when doing lubricant comparisons.

The presence of a relatively large amount of water vapor during dry pressing usually improves the action of lubricants, as has been found by several investigators.[69,70] Reference 69 reports that drying the

[65] P. Studt, Tribololgy Int'l., 22 (1989) 111 [see Chem. Abstr. 112 (1990) item 142439e].

[66] S. Watenabe, et al., Ind. Eng. Chem. Res., 28 (1989) 1264.

[67] V. Mody and D. J. Shanefield, unpublished results (1989).

[68] E. J. Motyl, Western Electric Engineer, 7 (March 1963) 3.

[69] S. M. Marschner, Ph.D. Thesis, Rutgers Univ., (May 1991) [available from UMI, Ann Arbor, Michigan; see particularly page 125] .

‡ See dagger footnote on page 89.

[70] J. S. Reed, "Introduction to the Principles of Ceramic Processing,"
J. Wiley, New York (1988) 343; R. S. Gates, et al., Tribol. Trans., 32 (1989) 357;
J. S. Reed, et al., Am. Ceram. Soc. Bul., 71 (1992) 105.

ceramic powder before placing it in the ball mill evidently removes labile water enough to allow the lubricant to become adsorbed more, because dry pressed green densities were higher in a statistical experiment when A-16SG alumina was baked prior to being put into the ball mill. Evidently water should be removed just before the lubricant is to be adsorbed, but later it should be present to act as an auxiliary lubricant.

It might seem anomalous that baking the powder to remove water vapor could have a beneficial effect, even when that powder is then to be put into liquid water in the ball mill. But, as mentioned earlier, our ancient ceramic processes turn out to be surprisingly complex, as an investigator looks deeper into the myriads of details. It seems possible that, either the liquid water is slow to become re-adsorbed (which is known to be the case with some "activated alumina" drying agents), or it is not water that is the main factor, after all, but something else is being done by the baking. For example, possibly the glycol "grinding aid" (see reference 16 of Chapter 5) that is adsorbed on Alcoa A-16SG alumina becomes chemically modified (cross-linked, etc.), or maybe it is removed by the baking step, in addition to the more obvious water vapor removal.

In cases where the shape of the final product is compatible with *pressure filtration* as a shaping process, the large amount of water present in the highly fluid slip provides excellent lubrication, and very high packing factors (thus high green densities) can be obtained. This was also discussed briefly on pages 96 and 233.

Looking toward the future, evidence from outside the field of ceramics indicated that liquid crystals (molecules that are oriented the same way in one dimension but not in all three) might be superlubricants with unusual effectiveness.[71] Recent work at Rutgers[72] with Alcoa A-16SG alumina, ball milling 1 hr. with 1 part per hundred (p.p.h.) aqueous ammonium polyacrylate (R. T. Vanderbilt Co., Darvan 821A, m.w. = 6,000), followed by adding 1 p.p.h. triethanolammonium oleate (triethanolamine plus oleic acid, dissolved in alcohol) and milling 1 hr., then pan drying and dry pressing at 6,000 psi, yielded an unfired packing factor of 75 vol. %, which is unusually high for fine powder.

---

[71] S. E. Friberg, et al., J. Phys. Chem. 88 (1984) 1045; G. Biresaw, "Tribology and the Liquid-Crystalline State," Amer. Chem. Soc., Washington, DC (1990).
[72] D. J. Shanefield and Al'Tereek M. Stembridge, unpublished work.

Excellent dispersants, if they are soft materials, are usually excellent lubricants. For example, fish oil, or polyethyleneimine ester, or stearyl alcohol (especially when heated with the powder to make a good chemical bond[73]), each lubricate ceramic powders almost as well as stearic acid does.

These same materials make good *release agents** for ease in getting dried green bodies out of molds or off supporting films, again providing they are soft. An example of the need for being a soft solid or a liquid is that sodium metaphosphate, a hard solid, does not provide lubrication or mold release. On the other hand, low m.w. polyacrylate, a very soft solid, does act as a lubricant and mold release agent.[74]

Defoamers (also known as *antifoams*) are surfactants which migrate to the air/liquid interface and lower the surface tension. This decreases the stability of bubbles. The defoamer must be so strongly driven toward the surface that it can displace other surfactants which might also migrate there but would increase the stability of the bubbles.[75] The competition for sites on the surface is not scientifically understood in detail, and the use of defoamers is largely trial and error. Suppliers typically offer a kit of about ten materials the user can try before choosing one to buy in larger quantities. The compounds are most often medium m.w. alcohols such as octanol, low m.w. silicones, fluorocarbons, and carbon-carbon triple bond materials ("alkynes") which are derivatives of acetylene, CHCH.

Biocides (also known as *fungicides* or *algicides*) are used to prevent algae growth in slip storage tanks, and these were mentioned in connection with cationic detergents (Fig. 8.4). An example is dibutylene trimethyl ammonium bromide, a quat salt that is used as a biocide. Mercury organometallic compounds such as phenylmercuric acetate were used before it was found that they are dangerous to the environment, but zinc organometallic compounds are used now instead. Zinc dimethyldithiocarbamate is an example. However, completely organic biocides can burn out with no residue. An example is hexahydro-1,3,5-triethyl triazine, which is a six-membered ring where every other atom in the ring is a carbon and every other one is a nitrogen.

---

[73] F. F. Lange, et al., J. Am. Ceram. Soc., 77 (1994) 922 [C_{18} alcohol,].

* A small amount of shrinkage during drying can aid in achieving mold release.

[74] Lubricants can aid in "green machining" (drilling, etc.) before firing.

[75] X. Tang, et al., J. Am. Ceram. Soc., 79 (1996) 510.

10. PROCESSING EXAMPLES

Each of the following examples is a concise "case study," in which typical formulations are presented, along with problems that often arise in practice, and also some expedients that the author has found useful in solving these problems. The conciseness was considered a virtue in writing this final chapter of the book, with the expectation that the heretofore-patient reader might appreciate a bit less verbosity at this point. However, an intensely interested reader might be on the lookout for some of the special details which have been tightly compressed into these notes, with the warning that "The Devil is in the details." For example, a few percent change in the amount of binder could cause a host of problems.

These formulations can not possibly fit all powders, and they are only meant to be starting points for further optimization, especially if the powder being used does not have the specific surface area shown in the listings. When starting from literature recommendations such as the present chapter, even if initial results are satisfactory, further optimization almost always requires a certain amount of trial and error. The field of ceramic processing is more complex than it might appear to be at first sight, and much of what the reader might desire to learn is at present unknown. Therefore the words "trial" and "error" are fully operative. The author must confess that a large component of success in devising optimized ceramic formulations seems to involve sheer good luck.‡ But a novice reader should take philosophical encouragement from two observations: (1) failures are not always the fault of the experimenter, and (2) many of the most difficult-seeming problems in ceramic processing do, indeed, get solved. Usually one's luck is enhanced by working overtime a few nights and weekends, mostly because the engineer can then go through more trials (and errors).

In spite of the above emphasis on "luck" and "error," the scientific principles presented in this book can still be useful in launching the

---

‡ Much of the author's experience in ceramics reminds him of a comment made by his compatriot in the U.S. Army while on night patrols during the Korean War in 1953: "I'd rather have luck than brains."

experimenter in the right general direction, and also in making some of the most important "midcourse corrections" based on at least a little bit of knowledge, rather than chance alone. In addition, much of the experience of other workers is available via the published literature. Each of the processes listed below, and most of the chemical additives, can be referenced further by judicious use of the general index of this book, although some of these items might have to be found by looking up a group of synonyms. The footnotes of the book can then provide access to further details. A great deal of other information, though widely scattered, is available through the free publications listed in Appendix III.

10.1 Dry Pressing[1]

	Weight	Volume
Alumina(* next page)	100 gm	100/4 = 25 cc (50 vol. %)
Ammonium Polyacrylate	1 gm	1 cc
Water	24 gm	24 cc
		50 cc total

Ball mill for 2 hrs. Then add the following:

Polyvinyl Alcohol	1 gm solids, predissolved in water
Glycerin	1 gm Plasticizer
Ammonium Stearate†	1 gm solids, plus water

(Dagger notes are on next page)

Ball mill for 1 hr. Then **spray dry**[2] at ≤300°C inlet air temp.

Sieve and collect only granules smaller than 100 mesh (149 μm), and larger than 325 mesh (44 μm). Do not let fines escape. A bag type dust filter or water spray scrubber after the cyclone separator are advisable (see T. Graffe, "Dust Collectors," Am. Ceram. Soc. Bul., 72 [1993] 69.)‡ (Dagger notes are on next page)

---

[1] B. J. McIntire, page 141 in S. Schneider, "Engineered Materials Handbook, Volume 4," ASM International (formerly American Society for Metals), Materials Park, OH (1990).
[2] P. J. Sherrington, et al., "Granulation," Heydeon Publishers, New York (1981); K. Masters, "Spray Drying," J. Wiley, New York (1985).

separator are advisable (see T. Graffe, "Dust Collectors," Am. Ceram. Soc. Bul., 72 [1993] 69.)‡ (Dagger notes are on next page)
If spray dried "granule" microspheres appear hollow when cut with razor blade or appear as toroidal rings under a microscope, the inlet air temperature is too high and the water is boiling instead of just evaporating. Lower the inlet temp., and if granules are then not fully dry, decrease their size by speeding up the disk in a centrifugal drier, or increasing the air in a two-fluid nozzle drier, or *increase solids loading*. Lowering surface tension[3a] with a nonionic surfactant can also make smaller, faster-drying droplets. (Instead of a spray drier, use a Rotovap®[3b] for uniformity while drying, then force through a 40 mesh sieve.)
Tumble in rotating pan and spray 1 gm H_2O (auxiliary plasticizer) into it as a fog. Sprinkle aluminum stearate powder or wax powder onto tumbling granules, as **external lubricant**, to prevent "lamination cracks" ("end-capping") during pressing. (In place of dry lubricant, spray aqueous emulsion of wax or stearic acid onto the tumbling powder.) "Age" the powder overnight for water penetration. **Press** into die. Punch diameter should be 0.003 inch smaller than die diameter, to allow air to escape. Hold vertical compressive force on body while ejecting, to further prevent lamination, by allowing time for rearrangement, thus minimizing springback.

Notes regarding previous page:
* Density is 4 gm/cc. B.E.T. specific surface area is roughly 4 m^2/gm. If a finer powder is used, proportionately more dispersant (and possibly more water) might be required, to be adjusted by trial and error.

† Internal lubricant. Alternatively, use emulsified wax (see pages 223 and 276).

‡ In a company making spray dried and dry pressed ferrite ceramics, the centrifugal spinner in the spray drier was running faster than usual, making very small droplets. These became very fine granules, which got through the cyclone separator and went out a chimney. A slight wind blew the granules onto the employee parking lot, where the black dust settled on windshields of cars. It had previously rained, the cars were wet, and the PVA redissolved. Then the sun came out and dried the ferrite onto the glass. After work, people trying to scrape off the black gritty deposits permanently scratched their windshields, including the company president. The engineer in charge of spray drying was summarily fired. (It might be noted here that the author was not involved in this particular incident.)

---

[3](a) K. J. Konsztowicz, et al., Ceram. Trans., 26 (1992) 46. (b) Rotating vacuum dryer, from Brinkmann Co., Westbury, NY, or from Buchi Ltd. in Europe.

10.2 Injection Molding*

Alumina**	100 gm	
Oleyl Alcohol	1 gm	Dispersant and Lubricant
Ethylenevinyl Acetate Copolymer	5 gm‡‡	Binder, Fluidizer when hot
Propylene Carbonate	1 gm	Plasticizer
Low M.W. Polyethylene Wax	10 gm	Binder, Fluidizer when hot
Hydrocarbon Oil	2 gm	Plasticizer

Mix in torque rheometer for 1 hr while being heated. (Alternatively, dissolve in hot toluene, ball mill, then evaporate the toluene.) Force into mold while being heated. If "laminations" occur, use more preheating, up to the point where the heat damages the binder, or the cooling time slows the process excessively. If firing causes sagging or warping or cracking, try increasing the solids loading, and/or packing tightly in grog.

Note: If flow is nonuniform, use funnel-shaped entrance passageway and opening ("sprue" and "gate"), and arrange for the flow to come into the longest side, rather than into the small end of the mold.

---

* I. I. Rubin, "Injection Molding," J. Wiley, New York (1972); B. Matsuddy, page 173.in "Engineered Materials Handbook, Vol. 4," Amer. Soc. for Mat'ls., Materials Park, OH (1991).

** Density is 4 gm/cc. B.E.T. specific surface area is roughly 4 m^2/gm.

‡‡ The copolymer is self-plasticizing (see S. K. Khanna, J. Audio Engr'ng. Soc., 25 [1977] 724), and self-deflocculating for powder ("accepts filler," in the plastics terminology," see inside front cover of IEEE Electrical Insulation Magazine [July 1992].)

10.3 Extrusion[4a]

Alumina	100 gm	
Ammonium Polyacrylate	1 gm	Dispersant
Hydroxypropyl Methyl Cellulose	3 gm	Binder
Glycerin	2 gm	Plasticizer
Ammonium Stearate	1 gm, plus water	Lubricant
Water	20 gm	

Mix in pug mill or torque rheometer for 15 minutes. Then apply vacuum to de-air while mixing. (Viscosity must be high enough for the wet green body to be self-supporting.) Force through the extrusion die. Heat while exiting die, to gel the binder. Blow hot air to keep gelled and evaporate the water.[4b] The body can be cooled when dry. (However, excessive thermal gelation can cause too much friction in the die.) Instead of using thermal gelation to cure sagging ("slumping"), the material can be partly flocculated to provide pseudoplastic flow. Partial support for sagging green bodies can be provided by a moving belt. Extrusion can also be done downward to prevent sagging.

If dilatancy occurs (can explode die!), the material then requires more liquid, or better powder particle size distribution, or better internal lubricant/dispersant system, or if all else fails, a lower extrusion rate. External lubricant (emulsified wax, etc.) can be injected onto inside wall of die through drilled passageways.

For tubes and honeycombs,[5] use the die as shown in Fig. 10.1.

---

[4](a) K. Laue, et al., "Extrusion," ASM Int'l., Materials Park, OH (1981); V. F. Janas, et al., Ceram. Eng. & Sci. Proc., 14 (1993) 370. (b) Microwave dielectric heating is an excellent way to provide additional heat in a uniform manner. See V. F. Janas, et al., Ceram. Trans., 36 (1993) 493.

[5] R. D. Bagley, U.S. Patent 3,790,654 (1974); I. M. Lachman, et al., U.S. Patent 3,885,977 (1975); R. D. Bagley, U.S. Patent 3,904,743 (1975) [Corning Co.].

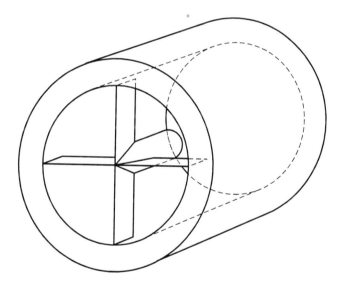

Fig. **10.1** A metal "spider" die, with central slug and four supporting vanes, used for extruding tubes. This drawing shows the view looking from inside the extruder.

10.4 Tape Casting[6,‡]

Aqueous

Alumina*	100 gm	
Ammonium Polyacrylate	1 gm	(More if higher B.E.T.)
Water	24 gm	
Ammonium Hydroxide	Adjust to pH of 8.2.	

Ball mill for 2 hrs. Then add the following:

Polyacrylate Emulsion	3 gm solids, plus emulsifying water	
Polyethylene Glycol, m.w.300	3 gm	Plasticizer

Ball mill for 1 hr.
De-air with vacuum until frothing suddenly increases.
Pump through filter.
Cast† on polyethylene terephthalate plastic film, at 6 in./min.,
using a doctor blade to control thickness.
Dry with heated air flowing opposite to casting direction.
For small batch, presaturate[7] the initial air with solvent vapor.

---

[6] D. J. Shanefield, in "Encyc. Mat'ls. Sci. and Eng.," M. B. Bever, ed., Pergamon Press, New York (1986) 4855 [general]; R. E. Mistler, Am. Ceram. Soc. Bul., 69 (1990) 1022; D. Hotza, et al., Mat'ls. Sci. & Engrng., A202 (1995) 206 [aqueous].

‡ For problem solving, see Table 10.1.

† For details of casting machine, see references 61 and 62 in Chapter 9.,

* Density is 4 gm/cc. B.E.T. specific surface area is roughly 4 m^2/gm. If a finer powder is used, proportionately more dispersant (and possibly more solvent) might be required, to be adjusted by trial and error. Also, finer powders might require up to 24 hrs. milling time for each stage (see ref. 61 in Chapter 9).

Nonaqueous

Alumina	100 gm	
Menhaden Fish Oil** (air treated)	1 gm	(More if higher B.E.T.)
Mineral Spirits	16 gm	
Isopropanol	8 gm	

Ball mill for 2 hrs. Then add the following:

Polyvinyl Butyral††	3 gm	
Dibutyl Phthalate	2 gm	Plasticizer
Polyethylene Glycol, m.w.300	2 gm	Plasticizer
(Stearic Acid [7]	2 gm	Additional plasticizer)

Ball mill for 1 hr. (See also asterisk footnote on previous page.)
De-air with vacuum until frothing suddenly increases.
Pump through **filter**.
Cast on polyethylene terephthalate plastic film, at 6 inches per minute, and dry with heated air flowing opposite to casting direction. (Additional heat can be applied via radiation from infra red heaters above, via conduction from resistance heaters mounted under the aluminum support plate, via dielectric heating within the cast slip itself, and by warming the slip just before casting.) Attach grounded, sharp wires near unrolling film to collect static electricity and prevent sparks. High humidity and ionized air blowers also tend to prevent sparks.
Dry tape can be "**calendered**" (squeezed between rollers) for improved compaction and reproducibility of firing shrinkage.

Notes:
** Regarding air treatment, see ref. 52b on page 242. Alternatively, try KD-1 or KD-3 polyimine esters, from ICI Co., Wilmington, DE (see "...imine" on pages 13, 231, and 244, and the following: E. DeLiso, and A. Bleier, in "Interfacial Phenom. Biotech. Mat'ls. Processing," Y. Attia, et al., eds., [1988] Elsevier, Amsterdam; E. Liden, et al., J. European Ceram. Soc., 7 [1991] 361). Another alternative is castor oil (see R. Moreno, et al., Am. Ceram. Soc. Bul., 74 [1995] 69).
†† Type B76 for coarse powder, B98 for submicron, (Monsanto, St. Louis, MO).

---

[7] To be used only for preventing cracking in difficult cases. See R. A. Desai and D. J. Shanefield, U.S. Patent 5,002,710 (1991) for more details.

Table 10.1 Tape Casting Problems And Their Solutions

PROBLEM	SOLUTION
Tape **sticks** excessively to carrier film	1. More release agent (dispersant if liquid, and/or plasticizer*), or use 2. Silicone surface-treated carrier film
Tape **releases** from carrier film too soon and curls up	1. Less of above, or 2. Add some more-powerful solvent (such as methylene chloride), to attack carrier film slightly
Tape is too **weak** to handle, once it is released from carrier film	1. More binder, up to the point of almost filling the pores between ceramic powder particles, or 2. Shorter milling time *after* adding the binder
Tape is hard but **brittle**	3. More plasticizer, up to the point of 2.5 times the binder weight (which would make the tape too sticky)
Cracks during *drying*	1. Higher solids loading in slip (might require up to 4 gm of dispersant per 100 gm of ceramic powder), or 2. More binder and plasticizer, up to the point of preventing high fired density 3. More release agent (dispersant or plasticizer) 4. Slower drying (lower temp. and/or air flow, or more highly saturated solvent vapor in the air)

(Continued on next page.)

* If excessive, can have opposite effect.

Table 10.1 Tape Casting Problems And Solutions (Continued)

PROBLEM	SOLUTION
Cracks during *firing* (and green density was too *low*)	1. Achieve a higher green density, by [a.] less binder and plasticizer, or [b.] better dispersion: higher quality dispersant, or optimized dispersant amount, or more milling time **before** binder
(and green density was *high* enough)	2. Slower firing heat-up (if cracked pieces *don't* match up) 3. Slower firing cool-down (if cracked pieces *do* match up)
Warps during firing	1. Try [a.] and [b.] above 2. Use *optimized* weight of *porous* pre-fired ceramic plates†
Fired density is too low (and *green* density was too low, and slip viscosity was <u>high</u>)	1. Try [a.] and [b.] above
(*green* density was too low, *but* slip viscosity was <u>low</u> enough)	2. Less binder and plasticizer, or 3. Optimize milling time **after** adding the binder and plasticizers

† Called "cover plates;" see page 420 of D. J. Shanefield and R. E. Mistler, Am. Ceram. Soc. Bul., 53 (1974) 416; also used later by others such as P. Nahass, et al., Ceram. Trans., 15 (1989) 355.

General Problem-Solving Sequence:
(1) Imagine all the reasonable *theories* that could explain what causes the problem.
(2) For each explanation in (1), imagine an *experiment* to prove (or disprove) it.
(3) Pick an experiment from the list in (2) that is fastest to do, and do it.
(4a) If (3) gives a positive result, do something based on that, to cure the problem.
(4b) If (3) gives a negative result, do the next-fastest experiment in (2), and so on.

Shanefield's Second Law:
"Don't do any experiment unless you can see in advance that it *could* lead to a conclusion, either theoretical or practical." (Many graduate theses ignore this.**)

** A drunk had dropped his keys in the dark but was looking for them under a light. When asked why, he said, "I can see here, but not over there where I lost them."

10.5 Slip Casting[8]

A wide range of compositions work well, including the following.

Clay, Silica, Feldspar*	130 gm‡	
Sodium Lignosulfonate†	2 gm	
Water	50 gm	(More, if necessary)
Calcium nitrate (if necessary)	0.2 gm	Flocculant for increasing casting speed if it is too slow[9]

> Pour slip into dry, porous mold (plaster or emulsion-polymerized plastic [for information on the latter, see ref. 1a in Chapter 9].). If thin-wall body is desired, pour excess liquid slip out of mold after the required thickness has deposited. Allow further drying and then separate mold halves.

Notes:

* Density is approximately 2.6 gm/cc. B.E.T. specific surface area is roughly 2 m^2/gm.

‡ 13 gm of this can be substituted by ball clay, in which case the organics are not needed.

† Alternatively, Na polyacrylate, or Na silicate. (Reference: A. K. Bougher, Am. Ceram. Soc. Bul, 73 [1994] 63.)

---

[8] C. H Schilling, page 141 in S. Schneider, "Engineered Materials Handbook, Vol. 4," ASM International (formerly American Society for Metals), Materials Park, OH (1990).

[9] I. Tsao and R. A. Haber, Ceram. Trans., 26 (1992) 73.

APPENDIX I

GLOSSARY OF CERAMICS WORDS

(See also the general index of this book.)

Ball clay: A type of clay with both small particle size (median approximately[1] 1 μm) and a naturally-occurring organic chemical residue of decaying leaves, etc. (called[2] *humic acid* or *lignite,* roughly 0.5% content[3]). Both the small particle size and the organic content contribute to good *plastic* flow when wet. (See also the word *plastic,* below.)

Ball mill: A large cylindrical container, usually ceramic, which is closed and turned on its side and then rotated on a pair of rubber-coated rollers. Sometimes it is made of metal, fitted with a water-cooling jacket, and lined with either polyurethane rubber or with close-fitting, curved ceramic bricks. For laboratory use, small polyethylene jars are sometimes used. The mill is 50% filled with small ceramic balls or cylinders (the "grinding media"). Slip is poured into the interstices between the balls, just filling up to the top layer of balls. The mill is rotated at 50% of the "critical speed," which is a higher speed at which centrifugal force would hold the balls against the inside wall of the mill and keep them from tumbling. At the 50% speed, tumbling is optimized, and both deagglomeration and mixing are achieved at high efficiency. (Larger diameters and harder surfaces require less time, because they are more energetic.) Newer variations of this machine include the "attriter," in which the mill is stationary and a paddle is rotated inside it, and the "pebble mill," in which very small balls are forced in and out of the mill by an external pump.

---

[1] For size range see: G. W. Phelps and M. G. McLaren, page 211 in "Ceramic Processing Before Firing," G. Y. Onoda and L. L. Hench, eds, J. Wiley, New York (1978) [Fig. 17.3].

[2] For chem. analysis see: S. P. Parker, "McGraw-Hill Encyc. of Sci. and Tech.," Vol. 8 (1992) 544.

[3] For amount of organics see: G. W. Phelps, et al., "Rheology and Rheometry of Clay-Water Slips," Cyprus Co., Sandersville, GA (1980) [page 6 and Table 1.1].

Bat: A supporting mold, usually porous, which holds the green body's shape while its top edges are being *fettled*.

Binder: In ceramics and powder metallurgy, a material which is added to the composition in order to provide a *plastic* type of *flow* property. In some cases it is a *fine* clay (often *ball clay*) or other inorganic material which remains in the overall composition after firing; in other cases (in fact, most cases), it is an organic chemical which is burned out during firing. Some ceramists use the word binder to mean all of the organics present in the green body, including the plasticizer, lubricant, etc. In any event, the main purpose of the binder is to provide strength to the material up until firing.

Bisque: Ceramic that has only been fired a short time or at a low temperature so that it is not fully dense.

Blunger: A mixer, usually for medium-viscosity slip, with a viscosity similar to that of paint or ink. (See also *pug mill.*)

Body: A ceramic object, unfired or fired.

Dunt (a verb): To crack during firing, sometimes because of cooling too fast. If the cracked halves can be fitted together perfectly, then cracking occurred during the cool-down part of firing, after sintering was completed. However, if the cracking had occurred during the first part of firing, the different random surfaces generated by the crack would not sinter equally, and the fired halves would then not fit together. The second type of problem is caused by uneven green density, etc., and slower cooling will not cure it.

Dust Pressing: See *Tempering.*

Encaustic: A coloring ink for decorating ceramics, usually a mixture of wax plus colored ceramic powder.

Engobe: To cover a *body* with an intermediate semi-glaze layer, helping to give it color (or whiteness), so that the true glaze layer which is later applied over it will have a more uniform color.

Fettle (a verb): To scrape off feather-edges, mold marks, or rough bottoms, etc. from the *green body.* Sometimes the fettling is done with a wet cloth.

Fine: Small (1 μm or less) particle size in a powder, or small grain size in a fired ceramic. (See also *ultrafine.*)

Flow: Motion of a material, relative to its surroundings.

> (a.) The most common type of flow involves either *shear* or else some other change in shape. Examples would be large amounts of material which are forced to move though a small hole, or around a sharp curve in a pipe.

> (b) Another type, less common, is *plug flow,* in which a volume of the fluid moves along fairly smoothly (relative to a pipe, etc.), without major changes in shape.

Fluid (adjective): Able to *flow* easily, usually when only forced by normal gravity. An example is that ice cream is not fluid (will not flow) at $0^{o}C$ but will flow at $25^{o}C$.

Fluid (noun): Usually a gas or liquid. However, a powder can be *fluidized* by being mixed with a certain amount of gas or liquid, and an example is the sand plus upwelling water in "quicksand," which acts like a liquid.

Forehearth: A long, refractory-lined trough which allows molten glass to cool slightly as it flows from the mixing furnace (typically at about $1300^{o}C$) to the molding machine (typically at about $1100^{o}C$).

Glost: Firing a glaze at a lower temperature than the bisque-fired body underneath had been fired. (This is different from "single fire," where both are at the same temperature.)

Gob: A large drop (typically a few inches in diameter) of hot, soft glass, to be pressed into a mold.

Green: Unfired but shaped ("formed") ceramic material, completely dried.

Grog: Broken-up, porous chunks of ceramic-type material, to be put around the green body during firing, so that the body is supported but will not stick excessively. (Sticking during firing is minimized with porous materials.)

Humic acid: See *ball clay.*

Jigger: A template that is held against a green body while the body is rotated, scraping off excess clay and thus giving the body its shape.

Jolleying: See *Pressure Casting.*

Kiln (usually pronounced "kill," without the n, although the "kiln" pronunciation is also correct.): A furnace for sintering.

Lehr: An annealing furnace for molded glass objects, to eliminate residual stress, typically holding the glass at about $530^{\circ}C$ for about 45 minutes. Also, a relatively low temperature furnace for firing the printing (or the decorating decals) onto glass bottles.

Lignite: See *ball clay.* (Note: "lignosulfonate" is a different material — see the main index of this book.)

Liquid phase sintering: (See also *sintering* below.) Densification of a green ceramic body, in which some of the main component of the mixture dissolves in a small amount of molten minor component (usually a eutectic), and it then recrystallizes on the surfaces of large grains. The large grains thus grow and fill in the pores. (This is probably the most common sintering mechanism occurring in industrial ceramics — in fact, completely solid state sintering seems to be a rarity — see glass intrusion experiment reported by D. J. Shanefield, in Advances in Ceramics, Vol 19, page 159, 1986.) In many cases, although it accounts for the major percentage of the overall firing shrinkage, it is followed by a small amount of final, *solid state sintering,* which primarily involves bulk diffusion.

Mineralizer: An inorganic additive such as lithium fluoride which enhances the calcining of a powder, causing the grains to grow larger and more dense. For example, when used during the calcining of aluminum sulfate, it causes fairly large, nonporous α-alumina particles to grow. If used in a compacted green body, this would promote *liquid phase sintering* and *vitrification.*

Muller: (sometimes called *mix muller*): An old style of mixer in which dry or damp powder is put on a horizontal plate, and a vertical wheel is rolled over it, going around repeatedly along a circular path. (This same type of machine was used to break wheat kernels down ["mill"] to smaller "flour" particles, hundreds of years ago, usually driven by oxen or waterwheels or windmills.) It is still widely used in ceramics factories, particularly to break down agglomerates in as-received powders.

Plastic (adjective): A type of *flow* property where the material is not *fluid* when only forced by normal gravity, but it becomes softer (more *fluid*) under the influence of greater force.

Plastic (noun): A polymeric organic chemical.

Pressure Casting, Molding, and Filtration: Shaping a green body by pressing very high viscosity damp clay (about 10% water) into a mold is usually called *pressure molding* or *ram pressing* or *jolleying*. If slightly more water is present and the mold is porous to absorb some of this water, the process is usually called *pressure casting* or *wet pressing*. If much more water is added, the resulting *slip* can be forced into a filter with most of this water passing through it and the clay remaining behind, and this process is usually called *pressure filtration*. The latter is capable of yielding extremely high green densities, because of the good lubricating action of the water (see References 27 and 32 in Chapter 8).

Pug mill: A stirrer for very viscous clay-plus-water mixtures, usually just before extruding. A vacuum is often applied inside, to de-air the mixture. Usually it involves a rotating helical screw ("auger") in a fixed metal cylinder. Sometimes the screw is made up from many individual short blades arranged in a helix around a simple cylindrical shaft.

Ram Pressing: See *Pressure Casting.*

Sagger: A *setter* (see below).

Setter: Previously-fired ceramic used to hold green bodies, supporting them as they are fired, and thus keeping them from "sagging" or "slumping." Sometimes it is simply a flat ceramic plate, and sometimes it is a complex array of refractory ceramic rods. Usually it is only *bisque*-fired, so that it is still porous and thus will be less likely to stick to the body being fired.

Shear: Parts of a material being in contact, some of the parts moving more than other parts, with the motions being in parallel to each other.

Sintering: To "fire" (heat at high temperature), slightly below the melting point of the main ceramic material, causing the material to shrink and thus densify. This could proceed via a *liquid phase* solution/dissolution process. Or it could proceed by *solid state* bulk diffusion in the main component of the material. Or it could (and usually does) proceed via both solid and liquid processes. (See also *liquid phase sintering.*)

Slip: Ceramic powder dispersed throughout a liquid. Other words for slip are *dispersion, slurry,* and *suspension.*

Tempering: With dry ceramic powders, this word means the addition of a few percent of water, in order to make the powder flow more easily in an extruder or dry pressing die. If a plasticized binder is present, about 2% water is usually needed, although more will be required for finer powders. If no binder is present, about 6% water is used, and this situation is sometimes called "dust pressing." With shaped glass bodies (and some ceramics that have glass in the grain boundaries), the word means to heat to the point where the material barely softens, thus allowing frozen-in stresses to become "relieved" by plastic flow, and the body is then cooled slowly and uniformly, in an essentially stress-free condition. With metals, the word means to heat and then suddenly cool, purposely building in stresses and thus hardening the material.

Ultrafine: Powder particle size less than about $0.1 \mu m$.

APPENDIX II

WORDS USED IN COLLOID SCIENCE

(See also the general index of this book.)

Absorption: One material penetrating into the open pores of another one. An example is mercury forced into the pores of firebrick by a porosimeter. The material being absorbed, such as mercury, is called the *absorbate.* The material being penetrated, like the brick, can be described by the adjective *absorbent.* (A less commonly used but still correct spelling of this word uses an "a," *absorbant.*)

Adsorption: One material on the surface of another. An example is nitrogen molecules attached to a cleaved single crystal surface. Any kind of physical or chemical bonding can be involved. (See also *sorption.*)

Aerosol: A *sol* composed of liquid or solid dispersed in air.

Agglomerate: Separate particles weakly sticking together.

Aggregate: Chunks of material strongly sticking together, usually by cohesion within one continuous phase.

Amphipathic: One part of a long molecule having a certain type of functionality (such as an acidic group) and another part of the same molecule having the opposite type of functionality (such as a basic group).

Catalyst: A material that can increase the speed of a chemical reaction without being permanently altered. Usually this is done by adsorption of the reacting chemicals, and a reaction pathway with a low activation energy is thus created.

Coagulation: Agglomeration from a previous sol. Modern authors use this term to mean the case where the particle is in the DLVO secondary minimum (see "DLVO" in main Index of book). Older literature calls that situation *flocculation.*

Colloid: A *sol* or a *gel.* (Also see those terms, listed below).

Deflocculant: A material that aids in the breaking down of *flocs*, that is, it aids in the act of *dispersion.* It can also be called a *dispersant..* It is usually a specialized type of soluble *surfactant*, although sometimes it is simply a very fine, insoluble powder.

Depletion flocculation: Large amounts of dissolved polymer causing flocculation of powder particles, even though there is not much adsorption.

Depletion stabilization: Large amounts of dissolved polymer causing *de*flocculation of powder particles, even though there is not much adsorption.

Dispersion: Discontinuous particles spread out within a continuous phase which is usually a fluid. (Same as *suspension* or *slurry*.) Can be either a true *sol* (that is, *stable*) or else unstable. (For comparison, see also *solution*.)

Emulsion: A *sol* composed of two liquids, although the word is often used loosely to mean a waxy solid dispersed in a liquid.

Fine particles: See *fine* in appendix I.

Floc: a large agglomerate with a very porous structure. In some cases where the particle sizes are extremely small, the meaning overlaps with that of *gel.*

Flocculant: A material that causes the formation of flocs. (*Flocculent:* An adjective, meaning flocculated.)

Gel: A colloid which behaves like a solid. In some cases this involves two *continuous* materials interpenetrating another, an example being solid but extremely porous silica glass, which has absorbed water. In other cases it is an agglomerate of *separate* particles in contact, plus a continuous fluid absorbed, and an example of this is ultrafine silica powder plus water. Lettuce and celery are gels (about 90% water), and many parts of the human body are gels.

Latex: A *sol* of polymer particles dispersed in water.

Lyophilic: A fine powder which is easily dispersed just by gently stirring it with a liquid. (From ancient Greek, "lyo-" means liquid, and "-philic" means loving.)

Lyophobic: A fine powder which must be mechanically forced into suspension by high shear stirring, usually requiring a dispersant.

Particle: See *solution.*

Peptized: A powder and liquid which have been made into a sol without mechanical action, either because the powder is lyophilic, or because a very powerful dispersant is used.

Powder: See *solution.*

Shear: See Appendix I.

Slippage: In viscosity measurements, the concept of a lubricated high degree of shear at the boundary between the moving liquid and the stationary container. When using a capillary viscometer, this could result in what chemical engineers call "plug flow," with little or no shear elsewhere in the liquid (see bottom of Fig. 6.11). Another example is in doctor blade tape casting (Fig. 4.2), where slippage is quite likely to occur at the bottom of the blade because of its small frictional area, resulting in low shear between the blade and the moving support film, although it is usually considered to be high within that small volume. There could be downward flow behind the blade, and moderate shear distributed within the liquid reservoir before the bottom of the blade. While slippage is not usually given much consideration by engineers, its possibility makes the estimation of true shear rates in ceramic processing difficult at best.

Sol: A discontinuous material dispersed within a continuous fluid as a *stable* suspension. Usually the dispersed material has a particle size[1] of less than approximately 0.1 μm (100 nm). Because of Brownian (thermally induced) motion, settling out will not take place in a gravity field of 1 g, so the dispersion is called *stable*. However, the sol particles can be settled out by a centrifuge. Also, they can be filtered out with a membrane "ultrafilter." Visible light is scattered by the particles of a sol, although they are not individually visible with an ordinary optical microscope.

Solution: More than one chemical species interspersed within the same phase, mixed on a molecular scale of sizes. In other words, individual molecules of the *solute* (the "discontinuous" material that has been "dissolved") are spread out within the *solvent* phase. These molecules are too small to be filtered out or settled out of the dispersion with ordinary equipment. (It should be noted that the phrase "too small" is somewhat arbitrary, since advanced techniques can sometimes be used get around these definitions.) For comparison, if the material that has been dispersed is in the form of larger units (usually groups of many molecules, but sometimes single very large polymer molecules), these large units are called *particles*. They might be *sol particles,* or they might be the still larger *powder particles* that can easily be settled out of a liquid.

Sorption: Absorption and adsorption occurring together, or indistinguishably.

Stable: Not changing with increases in time (see example under *sol*).

---

[1] In older literature, this size limit was considered to be 1 μm. However, there is no really precise size limit, because of the various possibilities of particle density, charge, adsorbed coatings, etc. In fact, modern technology can be used to force much larger particles into a *stable sol* form.

Surfactant (sometimes called a *wetting agent*): A material that tends to migrate to the surface of a liquid (either the liquid/gas or the liquid/solid surface) and change the surface energy relationships. Usually it lowers the liquid's surface energy, causing it to wet a solid more readily. Specialized types can act as *dispersants* (which are also called *deflocculants*).

Suspension: Particles spread out within a fluid. (Same as *dispersion.*) Can be either a true *sol* (that is, *stable*) or else unstable.

Ultrafine particles: See *fine* in appendix I.

APPENDIX III

INFORMATION SOURCES

(See also the general index of this book.)

Magazines and Other Publications

(Free to Professionals:)
Aldrich Chemical Co., Milwaukee, WI. [Catalog, with structural formulae of many thousands of chemicals, including some safety data, literature references, boiling points, and occasionally other useful data.]

Chemical Equipment, Morris Plains, NJ. [An all-advertisement magazine, with useful information about mixers, filters, etc., often including new designs.]

Ceramic Industry, Solon, OH. [Monthly magazine, with practical articles, market surveys, useful advertisements, etc.]

Environmental Protection, Waco, TX. [Monthly magazine. with practical articles, news about Government regulations, informative advertisements]

(Available in city public libraries:)
For listings of many other magazines and journals about each subject, see also "Ulrich's Int'l. Periodical Directory," R. R. Bowker Co., New Providence, NJ.

For a convenient method to find a book on an unusual subject such as "emulsion polymerization," or "polymers, water-soluble," etc., see also "Books in Print," R. R. Bowker Co., New Providence, NJ, updated yearly.

For listings of chemicals available from Japan, see "ChemIndex Japan," Tekno-Infor Corp., Louisville, KY, updated yearly.

Professional Societies *(Most of these issue technical newsletters.)*
American Chemical Society, Washington, DC.

American Ceramic Society, Westerville, OH.

American Institute of Chemical Engineers, New York, NY.

ASM International (formerly American Society for Metals), Materials Park, OH.

Chemical Manufacturers Association, Washington, DC.

Ceramic Manufacturers Association, Zanesville, OH.

Ceramic Society of Japan (Nippon Seramikkusu Kyokai), 22-17 Hyakunin-cho 2-chome, Shinjuku-ku, Tokyo 169, Japan.

Federation of European Materials Societies, Weinheim, Germany. (Includes various national ceramic societies, powder metallurgy, etc.)

Federation of Societies for Coatings Technology, Blue Bell, PA. (Includes paint.)

International Society for Hybrid Microelectronics, Reston, VA. (Cermet screen printing inks, ceramic multilayers, etc.)

Magnetic Materials Producers Association, Chicago, IL. (Powder metallurgy magnets and ferrite ceramics)

Metal Powder Industries Federation, Princeton, NJ. (Injection molded powder metallurgy, etc.)

Materials Research Society, Pittsburgh, PA. (Metals, ceramics, polymers)

The Materials Society, also known as ASM International (formerly American Society for Metals), Materials Park, OH. (Mostly metals, some ceramics and polymers)

National Association of Printing Ink Manufacturers, Hasbrouck Heights, NJ.

Porcelain Enamel Institute, Nashville, TN.

The Refractories Institute, Pittsburgh, PA.

Screen Printing Association International., Fairfax, VA.

Society of Glass and Ceramic Decorators, Washington, DC.

Tile Council of America, Princeton, NJ.

(For names and addresses of hundreds of other professional and manufacturers' organizations, see also P. K. Daniels, et al., "Encycl. of Associations," Gale Research Inc., Washington, DC [1994].)

Reference Books in Related Fields *(See also Chapter 4 for ceramics book listings, and other chapters for specialized texts such as Chapter 3 for organic chemistry, Chapter 6 for colloids, etc.)*

Cosmetics
E. W. Flick, "Cosmetic Formulations," Noyes Publications, Park Ridge, NJ (1992).

Ink
A. M. Wells, "Printing Inks," Noyes Publications, Park Ridge, NJ (1976). [Ink compositions in general.]

E. W. Flick, "Printing Ink Formulations," Noyes Publications, Park Ridge, NJ (1985).

D. J. Shanefield, "Electronic Thick Film Technology," page 284 in "Ceramic Films and Coatings," J. B. Wachtman and R. A. Haber, editors, Noyes Publications, Park Ridge, NJ (1993) [screen printing ink formulations for electronics].

Materials Science
M. B. Bever, "Encyclopedia of Materials Sci. and Engineering," Pergamon Press, New York (1986).

S. Schneider, "Engineered Materials Handbook," ASM International (formerly American Society for Metals), Materials Park, OH (1990). (See particularly Volume 4, "Ceramics and Glasses.")

Paint
T. C. Patton, "Paint Flow and Pigment Dispersion," J. Wiley, NY (1979).

G. P. A. Turner, "Paint Chemistry and Principles of Paint Technology," Chapman and Hall, New York (1988).

W. M. Morgans, "Outlines of Paint Technology," J. Wiley, NY (1990).

R. Woodbridge, "Principals of Paint Formulation," Routledge Divn., Chapman and Hall, New York (1991).

E. W. Flick, "Water-Based Paint Formulations," Noyes Publications, Park Ridge, NJ (1994).

Powder Metallurgy
F. V. Lenel, "Powder Metallurgy Principles and Applications," Metal Powder Industries Federation, Princeton, NJ (1980).

R. M. German, "Powder Metallurgy Science," Metal Powder Industries Federation, Princeton, NJ (1994).

Refractories (Firebricks)
Edwin Ruh, "Refractories for the Chemical Process Industries," The Refractories Institute, Pittsburgh (1984).

S. Carniglia et al., "Handbook of Industrial Refractories Technology," Noyes Publications, Park Ridge, NJ (1992).

Surfactants
P. Becher, "Emulsions," Reinhold, New York (1976) [HLB tables].

G. D. Parfitt, "Dispersion of Powders in Liquids," Elsevier, New York (1969).

D. H. Napper, "Polymeric Stabilization of Colloidal Dispersions," Academic Press, London (1983).

T. F. Tadros, "Surfactants," Acad. Press, New York (1984).

McCutcheon's Emulsions and Detergents," McCutcheon's Publ., Glenn Rock, NJ (1993) [technical data and supplier addresses for hundreds of surfactants].

R. B. McCay, "Technical Applications of Dispersants," Marcel Dekker, New York (1994).

Suppliers of Chemicals, Industrial Quantities
"Chem Sources, USA," Directories Publ. Co, Clemson, SC.

"Chem Sources, Europe," Directories Publ. Co, den Haag, Netherlands.

Chemical Buyers Direct Index, Walter De Gruyter Co., Berlin and New York.

[books containing lists of suppliers for thousands of chemicals, at lower prices than small-lot laboratory supplies].

Toxicity
N. I. Sax, et al., "Dangerous Properties of Industrial Materials," Van Nostrand Reinhold, New York (1993).

Publications that are difficult to obtain

The reader might have noted that, in selecting the literature references to be cited in this book from among the many other possible choices, emphasis was placed on those that are most readily available to ceramists, such as the American Ceramic Society publications, the New York Times, etc. However, other publications that are otherwise not available to the reader can almost all be obtained from the institution described below.

John Crerar Library, Univ. of Michigan, Ann Arbor, MI. [This library can send to local municipal public libraries the photocopies of almost any publication in the world. Although this is not a free service, it is subsidized by a foundation grant. It should be noted, however, that the interested reader might have to explain this service to some local librarians, since they are not always aware of it].

INDEX
(See also Appendices I and II.)

Acet- prefix, 34
Acetylene, 25, 62 footnote, 73, 279
Acids (see also pH),
  Lewis type, 12, 241, 270
  organic, 35, 51
Acrylates, 64, 258, 286
Activation energy, 102, 167, 181, 297
Addition reaction, 60
Additives,
  inorganic, 103, 228, 235, 258
  typical, 4, 5
  vs. impurities, 104
Adhesion, 245, 269
Adsorption, 105, 131, 203, 230, 270
  amount (weight), 133, 152, 239
  amount (layer thickness), 233, 237
  B.E.T., 126, 129
  competition, 134, 270, 271 footnote
  Langmuir, 134
  monolayer, 105, 134, 231
  multilayer, 103, 128, 134
  size-specific, 137
Advantages of organics, 3
Afterburners, 203, 263
Agar polysaccharide binder, 264
Agglomerates, 94, 109, 140, 267
Aggregates, 109, 125
Alcohols, 31, 209, 239, 253
Aldehydes, 33, 34, 51, 65, 71
Algae, Alginates, 44, 84, 86, 258 , 279
Alkanes, 26
Alkenes, 26
Alkyls, 26
Alkynes, see acetylene
Alpha position, 40
Alumina, 127, 192, 228, 278, 281
Aluminum nitride, hydration, 210
Amines and amides, 55
Amino acids, 55, 88, 89
Ammonia gas burnout, 250, 263
Ammonium ion, 9, 13
Amphipathics, 76, 138, 230, 244, 297
Amphoteric, 141

Angles of bonds, 21, 36, 47
Antifoams, 178, 279, 286
Aphorisms, 241, 275, 280, 289
Apparent & bulk density, 216, 273
Aromatic, 51, 209
Aryls, 49
Asbestos and cancer, 188, 189
ATSB, 32, 107
Auxiliary burners, 203, 263

Ball clay, 2, 291, 255
Ball milling, 94, 129, 232, 270, 291
Benzene, 49, 53
Base, see Lewis and pH
B.E.T. method, 103, 128, 129, 134
Binder, 4, 6, 244, 255
  amount, 268, 272, 288
  burnout, 203, 258, 261, 264
  migration, 256, 268, 272
Biocides, 279
Bis, 49
Block copolymer, 230
Bond
  angles, 21, 36, 47
  types, 8, 11, 48, 243
Branched chains, 42, 233, 252, 277
Bridging, 231, 235, 253, 272
Brownian motion, 153, 155, 253
Bubbles, 178, 225, 279, 286
Bunching of growth steps, 137
Burnout, 259
  catalytic, 203, 262, 269
  char (carbon residue), 259
  of binders, 203, 261, 264, 266
  of dispersants, 250
  oxidizing agents, 262
  reducing agents, 250, 263
Butane, 21, 41
Butter, 155

Calcining, 120, 125
Calendering, 287 (see also roll comp.)
Cancer, 53, 188
  de-listing, alumina as "cause," 192
  random background, 196
  rates unchanged, 44, 197
  skin, and ozone hole, 44

Capillary strength when wet, 274
Carbohydrates, 80
Carbon, residual, 259
Carbonyl compounds, 33
Carboxylic acids, 35, 51
Carboxymethyl cellulose (CMC), 83
Case studies, 280
Casting, see pressure, tape, slip
Catalysts, 66, 297
  in aiding burnout, 203, 262, 264
Cationic surfactants, 219, 221, 277
CAT-scan defect detection, 271
Cause not equal to correlation, 197,
    204, 205, 246
Cellosolve, 184
  toxicity, 195, 201
Cellulosics, 80, 163
Ceramacist vs. ceramist, 2
Ceramics, 1, 91
  books, 91, 305
  process complexity, 241
CFCs, 44
Chance, see luck, randomness
Char, 259
Charcoal adsorbent, 203
Charged powder particles, 140, 153
Chemical
  bonds, 8
  reactions, organic, 56
  reactions, powder, 209
Chlorocarbons, 27, 44
  natural, 196, 207
  nonflammable, 184
Churning to make butter, 155
Cis positions, 43
Citrate dispersant, 254
Clay, 2, 20, 132, 229, 291
  as a dispersant, 2, 235, 291
Closest packing,
  one size 117
  multisize, 118
Clustering of random data, 199
CMC, 83
Coagulation, 151, 155, 298
Colloids, 126, 297
Compaction, 93, 276
  from gel-shrinkage, 97, 271

Competition of adsorbates, 134, 270
Complexes, 47
Complexity of ceramics, 241
Condensation reaction, 60
Condensers for VOCs, 203, 266
Continuous porosity, 120 footnote
Coordinate bonds, 11, 48, 243
Copolymer, 76, 138, 230, 265, 283
Correlation not equal to cause, 197,
    204, 205, 246 footnote
Cost, 207, 258
Cover plates (porous weights), 289
CPFT particle size graph, 123
CPVC, 215
Cracks, 95, 254, 288
Cross linking, 20, 70, 75, 256
Curing of polymers, 66, 72
Critical powder volume conc., 215
Crystalline plastics, 19, 20, 69
CVD, 32, 47
Cyclic compounds, 36, 47,49

DDT dilemma, 205
Deagglomeration, 94, 140, 267, 286
De-air, 178, 225, 278, 284, 286, 295
Defects,
  causes, 94, 111, 254, 267
  avoidance, 177, 254, 288
  detection, 271
Deflocculant, 298 (see also dispersant)
Defoamers, 178, 279, 286
Degree of substitution (DS), 83
Delaney clause, 188
De-listing of alumina as a toxin, 192
Density,
  fired, 218, 289
  green, 115, 120, 215, 272, 278
  measurement, 120, 124, 215, 217
  tap, 124
  theoretical, 115, 120, 218
Depletion flocculation, 234
Desorption of water, 129, 270
Detergents, 218
Diameter, equiv. spher., 126, 127
Dielectric constant, 147, 238
Dilatancy, 164, 284
Dilemmas, 45, 205, 206

Dispersant, 6, 114, 211, 250, 298
  adsorbed thickness, 231, 233, 237
  amount adsorbed, 105, 133, 239, 253
  amount dissolved, 130, 212, 239, 253
  binder as, 244
  branched, 233, 277
  insoluble clay, 235
  molecular weight, 231, 238, 242,
    251, 253
  soft for steric hindrance, 238, 278
  solubility effect on amt., 133, 239
  solvent itself, 133, 238, 252
  ultrafine powders, 114, 250
  unsaturated, 244, 253
Dissolving criterion, 174, 176
Distribution of sizes, 120
DLVO curve, 144, 233
DNA, 89
Doctor blade tape casting, 95, 286
Donated electrons, 11, 12, 14
Double bonds, 13, 26, 245, see π
Double layer, 142
Dry ball milling, 129
Drying, 178
  accelerated, 254, 287
  gelation effect, 181
  the powder before using, 278
  shrinkage vs. solids loading, 92, 254
  spray drier, 281
Dry pressing, 100, 276, 281
DS (degree of substitution), 83
Dust-escape pollution, 189, 281
Dust pressing, 296

Earth warming effect, 204
Einstein equation, 165
Elasticity in slips, 161
Elastic stabilization, 155
Electron orbitals, 8
Electrophoretic mobility, 150
Electrostatic repulsion, 153, 158
Electrosteric, 158
Emulsion polymerization, 257
Emulsions, 218, 223, 235, 276, 298
Energy, see adhesion, bond types,
    free energy
Energy of activation, 102, 167, 181

Enthalpy, 157, 179
Entropy, 102, 156
  of mixing, 172
Environment, see afterburner,
    cancer, dust, fertility, greenhouse,
    halocarbons, Love Canal, nonionic,
    ozone, randomness, stillbirths,
    toxicity, VOCs
Epoxy, 72, 249, 256
Equiv. spherical diam., 126, 127
Esters, 46, 51, 56, 249
Ether, 41, 42, 83 (see also polyether)
Ethical dilemmas, 45, 193, 205, 206
Ethylene vinyl acetate copolymer
    (EVAC), 265, 283
Evaporation,
  of binder, 264
  of solvent, 178, 254, 281, 287
Examples, formulations, 280
Excess
  binder, 268, 272
  dispersant, 240
  milling, 232, 270
Explosive limits, 179, 184
External lubrication, 276, 282
Extrusion, 16, 83, 95, 259, 284

Fats, 58
Fertility effects, 191, 195, 205, 225
Filter, 267, 268 (see also pressure)
Firing, 93, 100, 107, 120, 288
Fish oil, 241, 243, 249, 287
Flammability, flash point, 179, 184
Flocculation, 146, 156, 167, 229,
    232, 234, 253, 290, 298
Flow, plastic, 1, 16, 85, 163
Fluorocarbons, 44, 73, 74
Foam, 178, 225, 278, 286
Food additives, 16, 46, 85, 163, 200
Forming, 7, 95
Formic acid, 65 footnote
Formulation examples, 280
Free
  energy, 103, 173
  radicals, 18, 42 , 61
  volume, 288
Freon, 44

Fungicides, 279
Funk and Dinger sizes, 125

Gaussian distribution, 121
Gelation, 16, 95, 151, 236, 284, 299
  and binder migration, 268, 272
  and drying rate, 181
  chemically induced, 86
  mechanically induced, 155
  thermally induced, 83, 258, 284
Gel-casting, 254
Gel shrinkage, 95, 97, 271
Glass transition temp., 78, 273
Global warming, 204
Gasoline, 42, 182, 208
Glycerin and glycols, 30, 273, 281
Glyceryl trioleate (sometimes called
    olein, and sometimes called
    triolein), 57, 58, 243
Glyceryl tristearate (sometimes
    called stearin, tri-stearin), 58, 243
Grain boundary amount,105
Green density, 113, 215, 217, 272
Greenhouse effect, 204
Green strength, 7, 258, 271
Group, organic chemical, 32
Gums, 86, 88, 226 footnote, 258

Heat of vaporization, 178
H-bond, 14, 89, 132, 168, 171
  adhesion, 131, 269
  agglomeration, 109
  energy, 179 table footnote
  viscosity, 167, 176
  gelation, 83, 284
  life is dependant on, 89
High solids, see solids loading
Hindrance, steric, 69
  aqueous, 226, 233, 237
  nonaqueous, 36, 42, 155
Homogenizer,
  chemical, 39, 275
  mechanical 224
Hormesis, 189
House of cards floc, 229
Humanity, end of, 205, 207
Humic acid, 2, 255, 291

Hydrated surfaces, 139, 209, 270, 278
Hydrocarbons, 24
Hydrogen bond, see H-bond
Hydrogen gas, wet, for firing, 262
Hysteresis, 163, 167

Imides, 55, 250, 251, 263
Imines, 55, 230, 244 (see also , KD-3)
Information sources, 302
Infrared spectra (IR), 17, 133, 277
Injection molding, 95, 256, 264, 283
Ink, 49, 85, 252, 304
Inorganic additives, 152, 103, 225
Inorganic dispersants, 225, 235
Iodine, 59, 189, 245
Isoelectric point (IEP), 151
Isomer, 30, 31, 41

KD-2 and KD-3 dispersants, 3, 13, 230,
    231, 244, 251, 287
Kelp, 86
Ketones, 33, 39
Kink sites on crystal face, 137

Lamination
  of multilayer tapes, 249, 275
  cracks, in dry pressing, 282
Langmuir adsorption, 134
Lecithin, 246
Lewis base, 12, 48, 132, 241, 270
Life and H-bonds, DNA, 89
Ligand, 49
Lignin and lignite, 2, 88, 255, 291
Lignosulfonate, 88, 229, 258
Linoleic acid and linolenic acid, 252
Linseed oil, 252
Liquid (see also solvent)
  crystal lubricant, 278
  phase sintering, 103, 294
Literature sources, 302, 307
Logic (see also correlation, dilemmas,
    paradoxes, randomness), 193, 197,
    280, 289
Lognormal distribution, 123
Loop train adsorbate, 231
Lot-to-lot change, 220 (see variability)
Love Canal, 197

Lubricant, 276, 278
  external, 276, 282
  internal, 276, 281, 283, 284
Luck, 196, 280 footnote

Mass action, dispersant, 240
MBI (methylene blue index), 130
Mechanism of reaction, 261, 266
MEK (methyl ethyl ketone), 33
Menhaden oil, 239, 241, 249, 287
Mer, 76
Meta positions, 40
Methyl cellulose, 16, 83, 163, 258
Migration of binder, 268, 272
Milling sequence, 232, 270
Mixing, 94, 173, 283
Moiety, 173
Molding, 96, 283
Molecule size, largest, 16
Molecular weight
  effects, see binder, dispersant,
    tangling
  number avg. vs. weight avg., 77
Monolayer adsorption, 105, 134, 231
Mooney-Celik equation, 165
Multicomponent mat'ls., 138, 241
Multilayer
  adsorption, 103, 128, 134
  lamination, 249, 275
Multiple-use additives, 4
Mylar®, 73

Names of organics, 20, 27
Natural
  halocarbons, 196, 207
  polymers, 80, 89
  versus synthetic, 4, 5, 88
NDE (or NDT), 271
New materials, safety tests, 193, 200
Nitrogen-containing organics, 55
Nitrogen oxides, see NO_x
NMP, 48
Nonaqueous, 4, 172, 176, 236, 287
Nondestructive evaluation (NDE),
    or nondestructive testing, 271
Non-DLVO, 233
Nonflammable solvents, 182, 184

Nonionic surfactants, 224, 282
  possibly a threat to life, 201, 225
Nonplastic, 2, 85
Nonpolar, 172, 176, 220 (see
    also polar)
Normal curve, 121
NO_x, 19, 203, 263

Octadecanoic acid, see stearic acid
Octadecanol, see stearyl alcohol
Octadecenoic acid, see oleic acid
Octadecenol, see oleyl alcohol
Odors, 51
Oil,
  fish, 241, 243, 249, 287
  motor, 69, 250, 251
  vegetable, 46, 58, 243, 252
Oleic acid, 35, 36, 46, 252
Olein (sometimes used to mean oleic
    acid, but also it sometimes means
    glyceryl trioleate), 57
Oleyl alcohol, 253, 283
Oligomer, 59, 138
Optimization, 112, 153, 253, 272, 288
Order of addition, see milling seq.
Organic chemicals,
  advantages, 3
  names, 20, 27
  reactions, 56, 61, 66
Organometallic, 47, 107 (see silane)
Ortho positions, 40
OSHA benefits not proven, 197
Outgassing,
  B.E.T., 129
  binders & precursors, 178, 266, 269
Oxidizing agents for burnout, 262
Ozone,
  atmospheric, 203, 263
  stratospheric, 19, 44

Packages, 63, 249, 254, 263, 275
Packing, closest,
  monosize, 116
  multisize, 118
Packing factor, 115, 217
Paint, 85, 99, 149, 163, 252, 255, 303
Paradoxes, 8 footnote, 205

Para positions, 40
Paraffin, 26
Particles, 113
  and surface area, 126
  size distributions, 110, 120
  ultrafine, dispersant for, 250
Parts per hundred gm (p.p.h.), 212, 278
PEG, see polyethylene glycol
pH effects, 141, 150, 234, 235
Phenolic compounds, 54, 71
Philosophy, 8, 196, 200, 241, 249, 275,
  280 (see also dilemma, paradoxes)
Phosphate, 228
  esters, 249
Phospholipids, 246
Photoresist, 63, 195
Phthalic acid compounds, 55, 73, 286
Physical bonds, 11
π bonds, 13, 50, 133, 172, 221, 243
Pinholes, 181
Plasticizers, 24, 79, 260, 273, 281, 288
  auxiliary, 282
Plastic flow, 1, 16, 85, 163
Plastics (polymers), 59, 249, 272
Plateau in curve, 109, 121, 129
Plug flow, 158, 160, 276, 293, 299
Polarity, 9, 17, 174, 220 (see also
  nonpolar)
Political activism, 204
Pollution, see environment
Polyacrylates, 63, 140, 213, 281
Polycarbonates, 74, 264
Polyelectrolytes, 151
Polyethers, 66, 226
Polyethylene compounds, 62, 73, 286
Polyethylene glycol, 66, 264, 287
Polyimines, see KD-3, imines
PMA, PMMA, 63, 140, 213, 281
Polymers, 59, 249, 257
Polymethacrylates, 63
Polysaccharide, 81, 264
Polyvinyl alcohol, 61, 274, 270, 281
Polyvinyl butyral (PVB), 65, 244, 270,
  274, 287 (see also residual OH)
  adhesion, 245, 270
  chemical properties, 264 (ref. 26)
Polyvinylidene difluoride (PVDF), 73

Polyvinyl polymers, 61, 62, 264
Polyvinyl pyrrolidone (PVP), 235, 264
Polyvinyls from acetylene, 62
Pores, 108, 120, 124, 215, 254
Porosity,
  calculation, 218
  open vs. closed, 108, 120 footnote
Powder, 113
  chemical reactions, 209
  drying by baking before use, 278
  metallurgy, 1, 75, 114, 255, 303, 305
  properties, 92
  ultrafine, dispersant for, 250
P.p.h. (parts per 100 gm), 212, 278
Precursors, 32, 107, 248, 268
Pressing, 96, 100, 276, 281
Pressure
  casting, 96, 177
  filtration, 96, 233, 278
Printing, 49, 85, 252, 304
Problem solving (see also logic), 289
Processing of ceramics, 91
Propoxy- group, nontoxicity, 202
Propylene carbonate, 74, 183, 264
Pseudoplastic, 163
PVA, see polyvinyl alcohol
PVB, see polyvinyl butyral
PVC (polyvinyl chloride), 63
PVDF (polyvinylidene difluoride), 73
PVP (polyvinyl pyrrolidone), 235, 264
Pyrrolidone ("pyrrolid*in*one"), 48

Quantum mechanics, 8, 11, 51
Quaternary salts, 11, 219, 224, 275

Radicals, 18, 26, 32, 42
Randomness, 121, 196, 280 footnote
Reactions, organic, 56, 61, 66
Reducing agent "burnout," 263
Reformulation, 126
Release agent, 274, 279, 288
Religions analogy, 205
Repulsion of particles, 153, 158, 246
Residual
  acetate and OH, 66, 245, 270
  carbon char, 259, 265
  solvent, 259

Resin, 72
Resonance, 11, 49, 51 footnote
Rheology, 158
Rifflers for powders, 128
Robust processing, 165, 254 (see also plateau, variability)
Roll compaction, 95
Rubber, 20, 70, 75, 78, 273

Saccharides, see polysaccharide
Sampling of powders, 128
Sand alleged to be a carcinogen, 188
Saturated organic compounds, 26
Schulze-Hardy rule, 145
Screen printing solvents, 180
Scrubber for waste gases, 266
Seaweed derivatives, 86
Secondary position, 31, 41
Sedimentation test, 211
Semisteric, 158, 238
Semisynthetic, 83, 85
Sexual functions, effects on, 19, 183, 199, 201, 205, 225
Shanefield's Laws, 275, 289
Shaping methods, 7, 95
Shrinkage, 94, 254, 268
   and solids loading, 92, 254
   calculation, 121, 218
Sieves, 121, 281
Silane derivatives, 32, 75, 248, 268
Silica powder, 163, 249 (see also sand)
Silicon nitride, 138, 250
Silicones, siloxane, silanol, 74, 75
Sinterability, 111, 114 footnote, 216
Sintering, 93, 100
   aids, inorganic, 104, 107
   liquid phase, 103, 294
Size distributions, 120
Skin cancer, 44
Skin formation, 181, 275
Slip casting, 98, 146, 290
Slippage, viscosity, 160, 276, 299
Slow drying solvents, 180
Smoke removal, 203, 263, 266, 269

Smoluchowski equation, 148
Soft dispersant, steric, 237
Sol-gel, 32, 184, 268
Solids loading, 92, 95, 135, 153, 176, 213, 228, 234, 251, 282
   and shrinkage, 95, 177, 254
Solubility, 173, 176
Solute, nearly-universal, 176
Solvents, 82, 171
   as dispersants, 133, 238, 253
   drying faster, 254, 287
   inexpensive organic, 208
   mixed, 240
   slow drying, 180
   thermal properties, 179
Sophisticated technologies, 49
Spectra, infrared, 17, 132
Sperm count, 205, 207
Spray drying, 281
Springback, see lamination cracks
Stabilized suspensions, 153
Starch, 4, 82, 258
Statistical mechanics, 19, 181
Stearic acid, 36, 155, 274, 277
Stearin, see glyceryl tristearate
Stearyl alcohol, 165, 252
Steps, on crystal faces, 137
Steric hindrance, 36, 42, 69, 155
   aqueous, 153, 223, 226, 233, 237
Stillbirths, 183, 199
Stories, 184, 194, 225, 246, 249, 270, 271, 280, 282
Strength,
   green, 7, 258, 271
   of plastics, 19, 272
Substrate, electronic, 73, 254, 178, 254
Succinic acid, succinimide, 55, 251
Sulfones, 74, 88, 227
Superlubricant, 278
Super-solute, 176
Suppliers, 306
Surface area, (see also adsorption)
   and size, 126
   measurement, see B.E.T.
   optimum, 112

Surface
  coverage, see adsorption
  energy, 101, 169
  steps and kinks, 137
  tension, 169, 282
Surfactants, 170, 218, 282, 301 (see
    also nonionic)
Suspensions, 153, 301

Tangling, 19, 24, 156, 231, 253, 272
Tap density, 124
Tape casting, 95, 166, 178, 275, 286
    and gel shrinkage, 97, 271
Teflon, 73
Temperature, 102, 153, 157, 167, 181
TEOS, 32, 184, 248
Terephthalic compounds, 55, 73, 286
Termination of polymerization, 61
Tertiary position, 41, 42, 252
Testing, see NDE, new materials
Tetrahedral bonds, 21, 36
Tetrakis, 49
Tg, 78, 273
Thermal gelation, 83, 184, 258, 284
Thermograv. analysis (TGA), 259, 268
Thermomech. analysis. (TMA), 78
Thermoplastic polymers, 62
Thermosetting polymers, 70, 256
Theta temperature, 153
Thickeners, see viscosifiers
Thixotropy, 163
Three roll mill, 94
Threshold of toxicity, 189
Toluene, 53, 172, 209, 286
  suspected poison, 194, 207
Toxicity, 53, 172, 189, 194, 207
Trans positions, 43, 46
Trichloroacetate super-solutes, 176
Triolein, see glyceryl trioleate
Tris-, 48
Tristearin, see glyceryl tristearate
Turpentine, 252

Ultrafine powders, dispersion of, 250
Ultrasonication, 110
Ultrasonic defect detection, 271

Unsaturation, see double bonds, 253
Unzipping binder, 265
Use first — test later, 193, 200

Vacancies, see free volume
Vacuum de-air, 286, 295
van der Waals bonds, 11, 21, 269, 270
Vapor vs. liquid volume ratio, 178
Variability, 61, 88, 95, 165, 220 (see
    also randomness, warping)
Vegetable oils, 46, 58, 243, 252
Vinyl polymers, 62
Vinylidene, 73
Viscoelasticity, 161
Viscosifiers, 16, 85, 163, 276, 284
Viscosity, 83, 158, 212
Vitrification, 108
VOCs, 203, 263
Volatility of solvents, 178, 182
Volume,
  bulk and apparent, 215
  free, 218
  of vapor vs. liquid, 178
  percentage example, 281

Warping, 98, 254, 268, 289, 296
Waste vapor, 203, 263, 266, 269
Water, (see also hydration)
  lubricant effect,277
  plasticizer effect, 274
  removal from powder, 278
  traces, 246, 249, 266
Wax, 12, 174, 221, 276
  emulsified, 223, 264, 282
Weak (mechanically) dispersants
  for steric hindrance, 237
Wet hydrogen gas, for firing, 262
Wet strength, 171, 255, 271, 274
Wetting, 168, 248, 276, 301
Wicking for binder removal, 266

X-ray defect detection, 271

Zero point of charge (ZPC), 151
Zeta potential, 150
Zwitterion, 230